Contents

Acknowledgments

Before all I thank my parents, Mary Ahern Barr and Donald Barr, to whom, humanly speaking, I owe everything. I thank my patient wife, Kathleen Whitney Barr, and our children, Thomas, Gregory, Victoria, Elizabeth, and Lucy, from whom I stole the time to write this book. My three brothers all played critical roles. William goaded and pressured me into actually writing the book. Hilary inspired me by his own literary endeavors. Christopher disagreed with me about a lot of things and forced me to sharpen or rethink my arguments.

I thank the Fellowship of Catholic Scholars. The opportunity to speak at their 1987 convention gave me the occasion to put some of my thoughts in order. I am greatly indebted to the journal *First Things*, and all the people connected with it, especially Fr. Richard John Neuhaus. Writing for *First Things* has been invaluable in many ways. Reviewing books for it gave me needed experience in writing and forced me to read numerous works that have been of use in preparing this book. I thank Professor Peter van Inwagen of the Philosophy Department of the University of Notre Dame for taking an interest in this project (despite disagreeing with some of my philosophical views) and helping to get the interest of the University of Notre Dame Press. I appreciate the efforts of the people at that press, in particular Jeff Gainey, Julie Beckwith, Ann Bromley, Christina Catanzarite, Rebecca DeBoer, John de Roo, Margaret Gloster, and Wendy McMillen, and of those who acted as readers for the press. My special thanks to Mark DePalma for help on the figures.

Finally, I wish to thank all those who have been willing to discuss with me questions concerning science and religion over the years.

The Conflict between Religion and Materialism

1 The Materialist Creed

In recent years there have been a great number of books written on science and religion. Some of these books claim that science has discredited religion, others that science has vindicated religion. Who is right? Are science and religion now friends or foes?

Perhaps they are neither. A frequently heard view is that science and religion have nothing to do with each other. They cannot contradict each other, it is said, because they "move on different tracks" and "talk about different realities." There is something to be said for this point of view. After all, there have been very few if any cases where scientifically provable facts have clashed with actual doctrines of Christianity or Judaism. Few believers any longer interpret the Book of Genesis in a narrowly literalistic way, and religious authorities no longer trespass on turf that belongs properly to scientists, as they did in the Galileo case. It is therefore perfectly true to say that science and religion are not in collision.

But while it is true, this viewpoint is too facile. The fact of the matter is that there *is* a bitter intellectual battle going on, and it is about real issues. However, the conflict is not between religion and science, it is between religion and materialism. Materialism is a philosophical opinion that is closely connected with science. It grew up alongside of science, and many people have a hard time distinguishing it from science. But it is not science. It is merely a philosophical opinion. And not all scientists share it by any means. In fact, there seem to be more scientists who are religious than who are materialists. Nevertheless, there are many, including very many scientists, who think that materialism is the scientific philosophy. The basic tenet of so-called "scientific materialism" is that nothing exists except matter, and that everything in the world must therefore be the result of the strict mathematical laws of physics and blind chance.

The debate between religion and materialism has been going on for a long time—for centuries, in fact. Why, then, the recent increase in interest in the subject? Is there really anything new to say about it? I think that there is.

1

What is new is that discoveries made in the last century in various fields have changed our picture of the world in fundamental ways. As a result, the balance has shifted in the debate between scientific materialism and religion. Many people sense this, but not everyone is exactly clear what these discoveries are or what they really imply. In this book I am going to give my own view of the matter, as someone who adheres to a traditional religion and who has also worked in some of the subfields of modern physics that are relevant to the materialism/religion debate.

Much of scientific materialism is based on certain trends in scientific discovery from the time of Galileo up to the early part of the twentieth century. It is easy to see why these trends led many thoughtful people to embrace a materialist philosophy. However, a number of discoveries in the twentieth century have led in surprising directions. Paradoxically, these discoveries, coming from the study of the material world itself, have given fresh reasons to disbelieve that matter is the only ultimate reality.

None of this is a matter of proofs. The discoveries of the earlier period did not prove materialism, and one should not look to more recent discoveries to prove religion. Even if religious tenets could be directly proven by science, the real grounds for religious belief are not to be found in telescopes or test tubes. Faith does not need to wait upon the latest laboratory research. What the debate is all about, as I shall explain later, is not proof but credibility.

I have said that the basic tenet of scientific materialism is that only matter exists. At that level, it is a very simple thing. On another level, however, scientific materialism is, like religion, a rather complex phenomenon. One can identify at least three highly interwoven strands in the materialist creed. In its crudest form it is a prejudice which looks upon all religion as a matter of primitive superstition, at best a source of some charming tales, like those of the gods of Olympus or Jonah and the whale, but at worst a dangerous form of obscurantism which breeds fanaticism and intolerance. Some religious beliefs, according to this view, may be more sophisticated than others, but none of them has any serious intellectual content.

Scientific materialism also comes in more refined philosophical forms. Here its critique of religion is essentially epistemological. It is acknowledged that there are religious ideas which have a certain intellectual appeal and internal consistency, but they are rejected as being unsupported by evidence. The assertions made by religion, it is said, cannot be tested and therefore cannot be accepted by a person who wants to be guided by reason rather than wishful thinking.

Finally, there is the "scientific" part of scientific materialism, which argues that religion, however believable it may once have been, has now been discredited by science. According to this view, the world revealed by scientific discovery over the last few centuries simply does not look anything like what we were taught to believe by religion. It is this claim that I shall subject to critical scrutiny in this book. The question before us, then, is whether the actual discoveries of science have undercut the central claims of religion, specifically the great monotheistic religions of the Bible, Judaism and Christianity, or whether those discoveries have actually, in certain important respects, damaged the credibility of materialism.

Before taking up this question, which is the main subject of this book, I will spend one chapter looking at materialism in its cruder form of anti-religious prejudice. This means making a brief digression even before I begin. However, I think this is necessary. Centuries of anti-religious myth stand in the way of a fair discussion of the real issues and must be gotten out of the way.

2 Materialism as an Anti-Religious Mythology

To begin with I will state the anti-religious myth in words that I think are quite typical of a great deal of writing on the subject.[1]

"Religion is the fruit of ignorance. Ignorant people, because they do not know how the world really works or the true causes of things, have always had recourse to explanations based on mythical beings and occult forces. They attribute the unpredictability of nature to the whims of gods and spirits. One sees this in the ancient myths and legends of primitive peoples. For example, in Greek mythology, thunder and lightning were the weapons of Zeus, storms at sea were caused by the wrath of Poseidon, and volcanic activity was associated with the subterranean workshops of Hephaestos, from whose Roman name, Vulcan, the word *volcano* comes.

"But religion is not just simple ignorance. It is a form of pseudo-knowledge. True knowledge — which is to say scientific knowledge — is based on reason and experience, on testable hypotheses and repeatable experiments. Religious beliefs, on the other hand, are based on the authority of ancestors or holy men or sacred writings — in other words, on someone's say-so. The fundamental opposition, then, between science and religion is the conflict, inherent and unresolvable, between reason and dogma.

"The defining moment in the history of science was the confrontation between Galileo and the Roman Inquisition. In this episode science and religion stood revealed in their truest and purest colors. It was the decisive contest between the two approaches to the world, the scientific and the religious, and religion lost. Its defeat proved the hollowness of religious authority's claim to special knowledge about the world.

4

"Science is the rational approach to reality because it deals with things that can actually be observed. Its statements can be put to the test. Religion, by contrast, characteristically deals with entities—God, the soul, angels, devils, Heaven, and Hell—that are admitted to be invisible. Its statements, because untestable, must be 'taken on faith.' 'Faith' is nothing but the wholly arbitrary acceptance of statements for which there is no evidence, and is therefore the very antithesis of reason: it is believing without reason.

"As science has progressed, religious explanations have given way to scientific ones. No evidence of God or of the soul has been forthcoming. Rather, these fictitious entities have less and less room to hide. They were meant in the first place to fill the gaps in our knowledge of the physical world, and consequently they are being steadily and inevitably squeezed out as those gaps are systematically closed. Science is the realm of the known, while religion thrives on the 'unknown,' on the 'unexplainable,' and on 'mysteries'— in short, on the irrational."

It is not too hard to show that most of this fairly standard anti-religious caricature is based on misunderstandings and bad history. In the first place, it is important to emphasize that the biblical religions did not originate in pre-scientific attempts to explain natural phenomena through myth. In fact, the Bible shows almost no interest in natural phenomena. It is certainly true that biblical revelation, both Jewish and Christian, has as a central part of its message that the universe is a creation of God and reflects his infinite wisdom and power. However, the scriptural authors evince no concern with detailed questions of how or why things happen the way they do in the natural world. Their primary concern is with God's relationship to human beings, and with human beings' relationships to each other.

In other words, the religion of the Bible is not a nature religion. Indeed, one of the great contributions of the Bible, which helped clear the ground for the later emergence of science, was to desacralize and depersonalize the natural world. This is not to deny that the Bible is overwhelmingly supernatural in its outlook, but that supernaturalism is concentrated, so to speak, in a being who is *outside* of nature.[2] No more were the Sun or stars or oceans or forests the haunts of ghosts or gods, nor were they endowed with supernatural powers. They were mere things, creations of the one God.[3] It is not an accident that as traditional Christian belief has weakened in Western society in the last few decades there has been a recrudescence of belief in the "occult."

What is true of the Bible is also true of Jewish and Christian teaching since biblical times: it has been very little concerned with attempts to give religious explanations of natural phenomena. If one looks at authoritative statements of doctrine from the time of the early church fathers down through the Middle Ages and the Renaissance, one does not find pronouncements about botany, or zoology, or astronomy, or geology. For example, the most comprehensive statement of Catholic doctrine until just recently was the *Roman Catechism*, sometimes also called the *Catechism of the Council of Trent*, published in 1566, not long before the Galileo affair. There is nothing in the *Roman Catechism* pertaining to natural phenomena at all. The same is true of the doctrines of the other branches of Christianity, and of Judaism as well.

One place where theologians did concern themselves with the natural world was in interpreting the first chapter of the Book of Genesis, often called the Hexahemeron, meaning "the six days." Even here, however, the central doctrinal concern was not the details of how the world originated, but the fact that it was created. St. Thomas Aquinas summarized the mediaeval church's attitude toward the Book of Genesis as follows:

> With respect to the origin of the world, there is one point that is of the substance of the faith, viz. to know that it began by creation, on which all the authors in question are in agreement. But the manner and the order according to which creation took place concerns the faith only incidentally, in so far as it has been recorded in Scripture, and of these things the aforementioned authors, safeguarding the truth by their various interpretations, have reported different things.[4]

The authors to whom St. Thomas was referring were the fathers and theologians of the ancient church, and, indeed, their interpretations of the Hexahemeron varied widely. In the East, the theologians of Alexandria tended toward very allegorical and symbolic interpretations, while those of Antioch and Cappadocia tended toward strict literalism. In the West, the greatest of the fathers, St. Augustine (354–430), adopted a very non-literal approach. To take an important example, St. Augustine held that the "six days" of creation were not to be taken literally as a period of time or a temporal succession. He held, rather, that all things were produced simultaneously by God in a single instant and subsequently underwent some natural process of development. Much earlier, St. Clement (ca. 150–ca. 216), Origen (ca. 185–ca. 254), and other Alexandrians had held the same view.[5]

In commenting on this issue, St. Thomas Aquinas said that the idea of successive creation was "more common, and seems superficially to be more in accord with the letter [of Scripture]," but that St. Augustine's idea of simultaneous creation was "more conformed to reason," and therefore had his (St. Thomas's) preference.[6]

This statement of St. Thomas perfectly illustrates another important point, which is that the church has always sought to give empirical reason its due.

Never (even, as we shall see, in the Galileo case) has the church insisted upon interpretations of the Bible that conflicted with what could be demonstrated from reason and experience. In his *Summa Theologiae*, St. Thomas cites the teaching of St. Augustine on the principles which should be observed in interpreting Scripture: "Augustine teaches that two points should be kept in mind when resolving such questions. First, the truth of Scripture must be held inviolably. Second, when there are different ways of explaining a Scriptural text, no particular explanation should be held so rigidly that, if convincing arguments show it to be false, anyone dare to insist that it is still the definitive sense of the text."[7]

Indeed, St. Augustine was sometimes quite vehement on this subject, obviously provoked by statements of some of the less learned Christians of his day. In a famous passage in his book *De Genesi ad Litteram* (*On the Literal Meaning of Genesis*), he wrote:

> Usually even a non-Christian knows something about the earth, the heavens, and the other elements of this world, about the motion and orbit of the stars and even their size and relative positions, about the predictable eclipses of the sun and moon, the cycles of the years and seasons, about the kinds of animals, shrubs, stones, and so forth, and this knowledge he holds to as being certain from reason and experience. Now it is a disgraceful and dangerous thing for an infidel to hear a Christian, presumably giving the meaning of Holy Scripture, talking nonsense on these topics, and we should take all means to prevent such an embarrassing situation, in which people show up vast ignorance in a Christian and laugh it to scorn. . . . If they find a Christian mistaken in a field which they themselves know well and hear him maintaining his foolish opinions about our books, how are they going to believe our books in matters concerning the resurrection of the dead, the hope of eternal life, and the kingdom of heaven, when they think their pages are full of falsehoods on facts which they themselves have learnt from experience and the light of reason? Reckless and incompetent expounders of Holy Scripture bring untold trouble and sorrow on their wiser brethren, . . . to defend their utterly foolish and obviously untrue statements, they will try to call upon Holy Scripture, . . . although they understand neither what they say nor the things about which they make assertion.[8]

How then, given these very reasonable attitudes of such high authorities as Augustine and Aquinas, did the Catholic Church end up, in the early seventeenth century, condemning the scientific theories of Galileo? Part of the explanation, no doubt, lies with personal failings of the people involved, but it also had a lot to do with the agitated times in which Galileo lived. The church was

caught up at that time in the great conflict of Reformation and Counter Reformation. The central accusation leveled at the Catholic Church by the Protestant reformers was that her teachings and practices were a corruption of the
original pure gospel found in the Scriptures. The proper way to interpret scriptural passages thus became the major bone of contention. In order to guard
against Protestant ways of interpreting Scripture, the church laid down at the
Council of Trent certain principles of interpretation. These moderate and sensible rules ended up being tragically misapplied in the Galileo case. Ironically,
this had the effect of producing an exaggerated literalism that was a departure,
as we have seen, from the church's own ancient traditions of scriptural interpretation.[9]

Whatever the historical reasons for the fateful blunder, however, the Catholic Church, even at that darkest hour in her relations with science, did not reject
the idea that truths about the natural world could be known through reason,
observation, and experiment. Nor did she assert that genuine scientific proofs
must give way before literal interpretations of the Bible. The very head of the
Roman Inquisition, Cardinal Bellarmine, wrote the following memorable words
to a friend of Galileo's named Paolo Foscarini:

> If there were a real proof that the Sun is in the center of the universe, . . . and
> that the Sun does not go round the Earth but the Earth round the Sun, then
> we should have to proceed with great circumspection in explaining passages
> of Scripture which appear to teach the contrary, and rather admit that we did
> not understand them than declare an opinion to be false which is proved to
> be true. But, as for myself, I shall not believe that there are such proofs until
> they are shown to me.[10]

As a matter of fact, such a "real proof" was not possible in Galileo's and Bellarmine's time. (Galileo believed he had such proofs, but in fact his proofs were
wrong.) Bellarmine tried to avoid the conflict, but, unfortunately, he had died
by the time of Galileo's second encounter with the Roman authorities.

Whatever else can be said about this lamentable episode, the following is
true: the condemnation of Galileo, rather than typifying the church's attitude
toward science, was manifestly an anomaly. For while the Catholic Church has
never been afraid to condemn *theological* propositions—in its long history it has
anathematized many hundreds of them[11]—only in the single instance of Galileo
did the Catholic Church venture to condemn a scientific theory.[12] And even
in that case it refrained from doing so in its most solemn and formal way, which
would have been irrevocable.

The fact is that the attitude of the church has overwhelmingly been one of
friendliness to scientific inquiry. Long before Galileo, and continuing to the

present day, one can find examples in every century, not merely of church patronage of science, but of important scientific figures who were themselves monks, priests, and even bishops. It is worth mentioning some of the more outstanding examples.

Robert Grosseteste (ca. 1168–1253), bishop of Lincoln, was the founder of the "Oxford School," to which has been traced the beginning of the tradition of experimental physical science.[13] Thomas Bradwardine (1290–1349), who became archbishop of Canterbury, was one of the first people ever to write down an equation for a physical process.[14] Nicholas of Oresme (1323–1382), bishop of Lisieux, made major contributions to both mathematics and physics. He discovered how to combine exponents, and developed the use of graphs to represent mathematical functions and prove theorems about them. He showed that the apparent daily motion of the Sun about the earth could be satisfactorily explained by rotation of the earth on its axis. Oresme also made important attempts to give a quantitative description of accelerated motion, and played a major role in developing the physical concept of inertia. His work may have helped to pave the way for the ideas of Galileo and Newton.[15] Nicolas of Cusa (1401–1464), a cardinal and an important figure in mediaeval philosophy, speculated not merely that Earth was in motion, as Copernicus later suggested, but far more boldly that *all* bodies, including both Earth and the Sun, were in motion in an infinite universe which had *no* center.[16] (Oresme had had similar ideas.) The great Copernicus (1473–1543) was an ecclesiastic, being a canon of Frauenberg Cathedral. He was probably an ordained priest at the time of his death.[17]

Fr. Marin Mersenne (1588–1648) is a well-known figure in the history of mathematics, and a certain kind of prime number is named after him. Less well known is that he invented the afocal forms of the two-mirror telescope, fundamental to the modern theory of reflecting telescopes.[18]

The tradition of Jesuit astronomy is well known.[19] A Jesuit contemporary of Galileo, Fr. Christoph Scheiner (1573–1650), made important discoveries about sunspots and the Sun's rotation on its axis, and is credited with discovering sunspots independently of Galileo.[20] Fr. Francesco Grimaldi (1613–1653) was a pioneer in lunar cartography, and he gave to many of the features of the lunar landscape the names by which they are called today. His published discoveries on the refraction of light preceded Newton's and he discovered both the diffraction of light and the "destructive interference" of light.[21] Fr. Giovanni Riccioli (1598–1671) discovered the first "binary" or double star.[22] Probably the greatest Jesuit astronomer was Fr. Pietro Secchi (1818–1878), one of the founders of astrophysics. He developed the first spectral classification of stars, which is the basis of that still used today; invented the meteorograph; and was the first to understand that nebulae were clouds of gas.[23] Not all priest-astronomers were Jesuits, however. A case in point is Fr. Giuseppe Piazzi

(1746–1826), director of the Palermo Observatory, who discovered the first asteroid, Ceres, in 1801.[24]

Fr. Lazzaro Spallanzani (1729–1799) was one of the leading biologists of his day. He first interpreted the process of digestion, showing it to be a process of solution taking place by the action of gastric juices. He performed experiments that disproved the hypothesis of "spontaneous generation," and did important research on such varied matters as fertilization in animals, respiration, regeneration, and the senses of bats. Nor was his work confined to biology. He also helped lay the foundations of modern vulcanology and meteorology.[25] Gregor Mendel (1822–1884), an Austrian monk, is universally honored as "the father of genetics" for his discovery of the basic laws of heredity.[26]

Abbé Henri Breuil (1877–1962), who has been called "the father of prehistory," was one of the leading paleontologists in the world and for decades the foremost expert on prehistoric cave paintings.[27] Abbé Georges Lemaître (1894–1966) was one of the originators of the Big Bang Theory, along with Alexander Friedmann.[28] Fr. Julius A. Nieuwland (1878–1936), a chemistry professor at Notre Dame, was a co-developer of neoprene, the first synthetic rubber-like compound.[29]

One could also mention such significant figures in the history of mathematics as Fr. Francesco Cavalieri (1598–1647), whose ideas played a role in the development of calculus; Fr. Girolamo Saccheri (1667–1733), whose work led up to the discovery of non-Euclidean geometry; and Fr. Bernhard Bolzano (1781–1848), who helped to put the branch of mathematics called "analysis" on a rigorous footing and to clarify mathematical thinking about infinite quantities.[30]

Obviously, had the church been hostile to science and reason, or had religious faith been incompatible with the scientific temper of mind, so many ecclesiastical figures would not have been found making major scientific discoveries. (Because it was Catholic authorities who blundered in the Galileo affair, I have given only Catholic examples here. But non-Catholic clergymen have also made important contributions to science, from Joseph Priestley (1733–1804), the discoverer of the element oxygen, who was a Protestant minister, to John Polkinghorne, a distinguished particle physicist of our own day who became an Anglican clergyman.)

Many will be surprised to learn that so many Christian clergymen have contributed so importantly to scientific discovery. The name of Gregor Mendel, of course, is familiar to most people; and those who have studied astronomy will know at least of the role of the earlier Jesuit astronomers. But on the whole, even among scientists, the larger picture of the church's involvement with science is not well known. The one tragic episode of Galileo has therefore overshadowed all the rest and come to typify in the mind of the public, educated and uneducated alike, the relation of science and religion.

Even so, most scientific materialists would concede that to be personally religious is not to be personally hostile to science. Even if they do not always have a balanced view of the history, they do know that many of the great founders of modern science, including Copernicus, Galileo, Kepler, Newton, Ampère, Maxwell, and Kelvin, were deeply religious men. And those scientific materialists who are themselves scientists know religious believers among their own scientific colleagues. (They may have heard, too, of the recent survey which showed that roughly half of American scientists believe in a personal God who answers prayers.) However, while admitting that religious people can be and often are good scientists, the scientific materialist nevertheless is convinced that the religious outlook and the scientific outlook are fundamentally at odds. For religion involves dogma, faith, and mystery, all of which, the materialist thinks, are inimical to the scientific spirit.

In fact, the charge is not simply that dogma, faith, and mystery are unscientific, but that they are essentially contrary to reason itself. To accept a dogma, it is thought, is to put some proposition beyond the reach of reason, beyond discussion, beyond evidence, beyond curiosity or investigation.

This view of dogma as anti-rational is based on a fundamental misunderstanding of what religious dogmas are. It is thought that the basis of dogma is emotion. Consider the following passage from a recent book: "Nothing could be more antithetical to intellectual reform than an appeal *against* thoughtful scrutiny of our most hidebound mental habits—notions so 'obviously' true that we stopped thinking about them generations ago, and moved them into our hearts and bosoms."[31] The author here was not specifically discussing religious dogma, but his words well summarize what the word *dogma* means for many people. To a religious person, however, a dogma is not something that is embraced from mere hidebound habit or feeling or wishful thinking, rather it is understood to be a true proposition for which there is the best of all possible evidence, namely that its truth has been revealed by God.

The believer in religious dogmas accepts that there are *two* ways that a thing may be known to be true: either empirically, through observation, experience, and the "natural light of reason," or through divine revelation. Accepting the one does not mean rejecting the other. In fact, in our everyday life we recognize that our knowledge does have a double source: there is what we have learned for ourselves and what we have learned from the information of others, whether teachers, friends, books, or common knowledge. Indeed, a little reflection shows that what we have actually derived from our own direct observation of the world without relying upon the word of others is but a very tiny part of everything that we do know. For a person to accept as knowledge only what he had discovered and proved for himself from direct personal experience would put his knowledge at the level of the Stone Age.

Taking something on authority, then, is not in itself irrational. On the contrary, it would be irrational never to do so. The question is when we should take something on authority, and on what kind of authority, and how far we should trust it. In the case of religious dogma, the authority is said to be God, who, it is claimed, has revealed certain truths—primarily truths about himself—to human beings. Such a claim is not in itself contrary to reason, for it is certainly hypothetically possible that there is a God and that he has revealed himself to man.

On the other hand, reason would require that before accepting religious dogmas we must have some sufficient rational grounds for believing that there is in fact a God, and that he has indeed revealed himself to man, and that this revelation truly is to be found where it is claimed to be found. And, indeed, these requirements of reason have always been admitted by the monotheistic creeds of Judaism and Christianity.

It is true that some believers, finding it difficult to give a satisfactory account of why they believe, have fallen back on the idea that belief is simply something one chooses to do, that it is its own justification, that it is a blind "leap." This is the view called "fideism." However, it is not the view of the traditional faiths.

If we take what is perhaps the most dogmatic faith of all, Catholicism, we find that it utterly rejects fideism, condemning it as a serious religious error. The First Vatican Council, in 1870, made the following declaration:

> In order that our submission of faith be nevertheless in harmony with reason, God willed that exterior proofs of his revelation . . . should be joined to the interior helps of the Holy Spirit. . . . The assent of faith is by no means a blind impulse of the mind.[32]

The same council formally condemned the proposition that "people ought to be moved to faith solely by each one's inner experience or by personal inspiration."[33] Rather, the council emphasized that there are objective facts and arguments in favor of true religious belief. Indeed, the council declared that the existence of God could be known with certainty *without* faith and *without* divine revelation by the "natural light of human reason."[34]

It might be thought that Protestantism, with its doctrine of *sola fide* ("by faith alone"), its emphasis on the inner light, and its greater distrust of human natural powers, including the "natural light of reason," might embrace fideism. But that is incorrect. For example, in Calvin's monumental summa of Protestant belief, *Institutes of the Christian Religion*, one of the first chapters is entitled "Rational proofs to establish the belief of the Scripture."[35] Protestants, no less than Catholics, believe that faith in God is a "reasonable service," to use a scriptural phrase.

This is not to say, of course, that very many religious believers would be able to give a precisely elaborated account of the grounds for their beliefs, with every step rigorously justified in the manner of a mathematical proof. But, for that matter, how many people could give such an account for anything they believe? Very few. People are not, as a rule, that methodical or analytical. This does not mean that they do not have, at some level, implicit and not consciously formulated, good reasons for believing as they do. As G. K. Chesterton observed, "most people *have* a reason but cannot *give* a reason."

It must not be thought, however, that religious faith is simply a matter of proofs and evidence.[36] On the contrary, the certitude of faith is claimed to be itself a gift from God, a result of what the passage quoted above called "the interior helps of the Holy Spirit." But unless there were also "exterior proofs" of revelation, belief would not be "in harmony with reason."

The first article of religion, of course, which must be believed before any divine revelation can be accepted, is that there is a God. As we have seen, the Catholic Church claims that this can be known with certainty by the natural light of reason. And, indeed, throughout history theologians and philosophers have furnished a wide variety of arguments for the existence of God. The basic outlines of one such argument can be found in both the Old and New Testaments. In a famous passage in his Epistle to the Romans, St. Paul asserts that, although God is in himself invisible, his "eternal power and godhead" are "from the creation of the world . . . clearly seen, being understood from the things that are made."(Rom. 1:20) St. Paul is saying here that one may reason from the existence of an effect (in this case the existence of the world itself) to the existence of its cause (in this case God). The same argument is made by St. Irenaeus of Lyons, writing in the second century: "Creation itself reveals him that created it; and the work made is suggestive of him that made it; and the world manifests him that arranged it."[37] Calvin in his *Institutes* puts the same thought in these words: "God [has] manifested himself in the formation of every part of the world, and daily presents himself to public view, in such manner, that they cannot open their eyes without being constrained to behold him."[38]

This basic line of argument has been refined and developed in several forms, which go by such names as the Cosmological Argument, the Argument from Contingency, and the Argument from Design. There are also arguments for the existence of God based on the existence of an objective moral order and on the nature of truth and our capacity to know it.

Beyond grounds for believing that God exists, there must be some grounds for believing that there has been a divine revelation if dogmatic religious belief is to be truly rational. Again, both Christians and Jews have an array of arguments, largely historical in nature, for the fact of revelation.

It is beyond the scope of this book to present or develop either the philosophical or the historical arguments in detail. (However, the Argument from Design is discussed at length in chapters 9 through 13, and certain of the other philosophical arguments are briefly discussed in appendix A, where I attempt to explain some basic, traditional ideas about God and Creation.) The point that I wish to emphasize here is simply that Jewish and Christian thought have taken very seriously the importance of evidence and argument and the necessity of rational grounds for belief.

One of the common complaints against religious dogma is that it is a substitute for rational inquiry, that it puts an end to thought. However, this is not the actual experience of the Jew or Christian, for whom revealed truths are a source of light in which new things may be seen and new insights arrived at. This is well expressed by the motto of my alma mater, Columbia University: *In lumine tuo, videbimus lumen.* ("In Thy light, shall we see light.") The ideal of Christian theology has always been summed up in St. Augustine's phrase *"fides quaerens intellectum,"* "faith seeking understanding," or as he also expressed it, "I believe in order that I may understand." (*"Credo ut intelligam."*)

The misconception that faith is opposed to rational inquiry seems to have a lot to do with the word *mystery*. In the writings of materialists, as we shall see later, the words *mystery, mystery-mongering,* and *mysterianism* crop up repeatedly as bugaboos to be avoided at all costs. The idea appears to be that a mystery is something dark and off-limits, an aspect of reality that is essentially irrational and unintelligible. A mystery is thought to be, one might say, the dark shadow cast by a dogma. This is not only a misconception, but really the opposite of what a religious mystery is to a religious person. Dogmas do not shut off thought, like a wall. Rather they open to the mind vistas that are too deep and broad for our vision. A mystery is what cannot be seen, not because there is a barrier across our field of vision, but because the horizon is so far away. It is a statement not of limits, but of limitlessness.

A religious "mystery" is not a statement that reality is in itself unintelligible. On the contrary, belief in God is bound up with the idea that reality is *completely* rational and intelligible. This is akin to the scientist's faith that his own questions about the natural world have rational and intelligible answers. This attitude of the scientist is *also* a form of faith, for the scientist is convinced in advance that the intelligible answer exists, even though he is not yet in possession of it. The fact that he is searching for the answer is proof that he does not have it, but it also attests to his unconquerable conviction that the answer, though presently not in sight, exists. The scientist knows that there is some insight, some act of understanding, which he currently lacks, that would satisfy the rational mind on the particular point he is investigating. The religious believer's faith is an extension of this attitude: he knows that there is some insight, some act of understanding, that would constitute complete intellectual satiety, because it would be a state of complete understanding of reality. However, he realizes, being sane, that such a state of perfect understanding is not achievable by a finite mind such as his own. Rather that insight, that act of understanding, is the state of being of a perfect and infinite mind, namely God—it is, indeed, what God *is*. In the words of the Jesuit philosopher Bernard J. F. Lonergan, God is the "unrestricted act of understanding."[39]

So, the complete intelligibility and rationality of reality corresponds to the existence of a supreme intelligence, a supreme reason. The last person, therefore, who would say that there is anything about the world created by God that is inherently irrational is the Jew or Christian. However, the Jew or Christian knows that he is not himself God, and therefore will never be in a state of perfect understanding about all of reality. And, in particular, he knows that he can never comprehend God as He is in Himself, since God is an infinite mind. As stated earlier, the dogmas of faith concern primarily the nature of God. And it is for that reason that they are mysterious—not because they are not intelligible in themselves, but because they are not intelligible completely *to us*. They are of course intelligible *to God,* who comprehends completely all that is real, and therefore completely comprehends his own nature.

The reason that there are mysteries is that God is infinite and our intellects are finite. Thus the divine nature is not "proportionate" to our minds, as the mediaeval theologians would put it. However, the natures of things in the physical world are certainly finite, and therefore *are* proportionate to our intellects. There is consequently no reason whatsoever that comes from Jewish or Christian belief to have any doubt that we are capable of understanding the physical world. It is for that reason religious mystery hardly touches at all upon the matters which the physicist studies. Thus the idea that religion, because it acknowledges mystery, must be the enemy of natural science is unfounded.

Now, while religious dogmas do not in fact limit the kinds of things one is able to think about, materialism obviously does. The materialist will not allow himself to contemplate the possibility that anything whatever might exist that is not completely describable by physics. That is simply a forbidden thought. It is usually not even felt to be necessary to argue against it. Admittedly, many materialists will say that forbidding one to speak of non-material entities is simply a matter of scientific "methodology." Natural science investigates matter, they say, and so anything that might go beyond matter is outside of scientific discussion. However, it is hard to see why this should be so. For example, one can imagine investigating human psychology in a perfectly scientific way without prejudging whether the human mind is entirely explicable in terms of material processes. In any event, for most materialists it is not really only a question of methodology. The non-material is considered simply beyond the pale of rational discourse. In short, the materialist's notion of what a dogma is, though quite unfair to religious dogma, exactly fits his own views.

One sees this materialist dogmatism displayed in every field of inquiry from the philosophy of mind, to artificial intelligence, to psychology, to biology. For example, the former editor of *Nature*, Sir John Maddox, in his book *What Remains to Be Discovered*, describes the immense complexity of the human brain and shows how little we yet know about its neural circuitry and detailed functioning. And yet he feels entitled to conclude, "An explanation of the mind like that of the brain *must* ultimately be an explanation in terms of the way that neurons function. After all, there is nothing else on which to rest an explanation."[40] [emphasis mine] One of the facts that is most difficult for materialism to deal with is consciousness. (Though, since it is more a philosophical problem than an issue in physics, I do not discuss it at length in this book.) The philosopher David Chalmers, in his book *The Conscious Mind*, summarizes the various materialist approaches to the problem. One is what he calls "don't-have-a-clue materialism," which he defines as the following view: "I don't have a clue about consciousness. It seems utterly mysterious to me. But it must be physical, *as materialism must be true*."[41] [emphasis mine] Such a view, he finds, "is held widely, but rarely in print." Materialists regard consciousness as at most a merely "passive" by-product of physical processes in the brain. In surveying current thinking about consciousness, the scientist and philosopher Avshalom Elitzur concluded (disapprovingly), "I think one may talk here about the dogma of passivity."[42]

What is most puzzling to the religious person about this materialist dogmatism is its lack of foundation. The religious dogmatist, after all, accepts certain truths as dogmas only because he believes them to have been revealed by God. But the materialist obviously cannot claim divine authority for his statement that only matter exists. On what basis, then, does the materialist's apodictic certainty rest?

Is materialism claimed to be *self-evidently* true? If anything would appear to be self-evident it would be that there are certain things, such as ideas, concepts, and minds, which are of a different sort than material objects. If materialism were self-evidently true, one might expect it to be the common view of ordinary people, and obviously it isn't. Is materialism definitively proven by philosophical or scientific demonstration? Are there no respectable arguments that could bring its truth into doubt in the mind of an intelligent person? If so, then how can one explain the large number of philosophers and scientists who disbelieve in materialism and bring forward arguments drawn from many considerations against it? We shall meet some of these people and some of their arguments later in this book.

As we examine some of the arguments for materialism later, we shall see that ultimately all of them are completely circular. They all seem to boil down in the end to "materialism is true, because materialism *must* be true." The fact seems to be that the philosophy of materialism is completely fideistic in character.

Not only is materialism as it is usually encountered more fideistic than the faith of the ordinary religious believer, it is also far more narrow and intellectually confining. That is because it is essentially a negative proposition. A person who believes that there is something about the human mind that goes beyond matter has a great deal of freedom of thought in this area. How much of the human mind can be physically explained is for him an open question. He may think (like David Chalmers) that all of human behavior is entirely explicable in physical terms but that human subjective experience is not. Or he may believe (like Avshalom Elitzur) that certain aspects of human behavior also go beyond physical explanation. He may (like Chalmers) think that sensation involves a non-material aspect of the mind, or he may believe (with Aristotle) that only certain functions of the human intellect do. He may believe that some ultimate theory will encompass in one scientific framework both material and non-material realities, or he may believe that the divide between matter and spirit is fundamental.

The materialist, by contrast, is in a straitjacket of his own devising. *Nothing* is allowed by him to be beyond explanation in terms of matter and the mathematical laws that it obeys. If, therefore, he comes across some phenomenon that is hard to account for in materialist terms, he often ends up by denying its very existence. For instance, many materialist philosophers deny that there really is any such thing as subjective experience. Philosophers call this view "eliminativism." What cannot be explained by the theory is eliminated from consideration. Some renowned philosophers, such as W. V. O. Quine, have denied that there are any mental experiences or events at all. Quine says that the existence of mental or conscious processes must be "repudiated."[43] As we shall see later, there are many thinkers who, in order to escape certain anti-materialist arguments that are based on human rational powers, are willing to abandon, in effect, a belief in human rationality—including, of course, their own. Almost all materialists deny that free will exists; they deem it an "illusion." And so it goes. Anything that stands in the way of materialism is ignored or denied. The materialist lives in a very small world, intellectually speaking. It is a universe of huge physical dimensions, but very narrow, for all that. There is no purpose in this universe. Even human acts are entirely determined by physical processes. Just as the astrologer believes that his life is controlled by the orbits of the planets, the materialist believes that his own actions and thoughts are controlled by the orbits of the electrons in his brain. Our moral or aesthetic judgments are, in the final analysis, just emotional reactions, just chemistry. Even our very "selves" are just convenient fictions; there is no real unitary self that stands behind the welter of images, impulses, drives, and thoughts flickering through our neural circuitry.[44]

The believing Jew or Christian does not feel the need to be embarrassed when materialists attack religion as "anti-scientific" or irrational. For he regards his own beliefs as not less but far *more* rational than those of the materialist. He regards them as providing a fuller, more coherent, and more sensible picture

3 Scientific Materialism and Nature

Though I have spent some time discussing them, it is not the historical prejudices of some scientific materialists that are the main subject of this book, but rather the interpretation given by materialists to what science has actually discovered in the last four centuries about the natural world. To use the political language of our day, what I wish to discuss is not the "spin" which some materialists have put on religion or history, but the spin they have put on scientific facts and theories.

Passing, then, beyond the bias and bad history which often accompanies it, one finds that scientific materialism has a case to make against religion that is based upon the discoveries of science itself. Again, let me state this case in words that might be used by a typical materialist:

The Scientific Materialist's View of Nature

"The world revealed by science bears little resemblance to the world as it was portrayed by religion. Judaism and Christianity taught that the world was created by God, and that things therefore have a purpose and meaning, aside from the purposes and meanings we choose to give them. Moreover, human beings were supposed to be central to that cosmic purpose. These comforting beliefs can no longer be maintained in the face of scientific discoveries.

"The universe more and more appears to be a vast, cold, blind, and purposeless machine. For a while it appeared that some things might escape the iron grip of science and its laws—perhaps Life or Mind. But the processes of life are now known to be just chemical reactions, involving the same elements and the same basic physical laws that govern the behavior of all matter. The mind itself is, according to the overwhelming consensus of cognitive scientists, completely explicable as the performance of the biochemical computer called the brain.[1] There is nothing in principle that a mind does which an artificial machine could not do just as well or even better. Already, one of the greatest creative chess geniuses of all time has been thrashed by a mass of silicon circuitry.[2]

"There is no evidence of a spiritual realm, or that God or souls are real. In fact, even if there did exist anything of a spiritual nature, it could have no influence on the visible world, because the material world is a closed system of physical cause and effect. Nothing external to it could affect its operations without violating the precise mathematical relationships imposed by the laws of physics. The physical world is 'causally closed,' that is, closed off to any non-physical influence.

"All, therefore, is matter: atoms in ceaseless, aimless motion. In the words of Democritus, everything consists of 'atoms and the void.'[3] Because the ultimate reality is matter, there cannot be any cosmic purpose or meaning, for atoms have no purposes or goals.

"Once upon a time, scientists believed that even inanimate objects did have purposes or goals: 'ends' which they sought or toward which they tended. For example, heavy objects were said to fall because they sought their proper place at the center of the earth. That was the idea in Aristotelian physics. It was precisely when these ideas were overthrown four hundred years ago that the Scientific Revolution took off. With Galileo and Newton, science definitively rejected 'teleology' in favor of 'mechanism.' That is, science no longer explains phenomena in terms of natural purposes, but in terms of impersonal and undirected mechanisms. And, of course, if there are no purposes anywhere in nature, then there can be no purpose for the existence of the human race. The human race can no longer be thought of as 'central' to a purpose that does not exist.

"Science has dethroned man.[4] Far from being the center of things, he is now seen to be a very peripheral figure indeed. Every great scientific revolution has further trivialized him and pushed him to the margins. Copernicus removed Earth from the center of the solar system. Modern astronomy has shown that the solar system itself is on the edge of a quite ordinary galaxy, which contains a hundred billion other stars. That galaxy is, in turn, one of billions and perhaps even an infinite number of galaxies. Earth is an insignificant speck in the vastness of space: its mass compared to all the matter in the observable universe is less than that of a raindrop compared to all the water in all the oceans of the world.[5] All of recorded human history is a fleeting moment in the eons of cosmic time. Even on this cozy planet, which we think of as ours, we are latecomers. *Homo sapiens* has been around at most a few hundred thousand years, compared to the 4 billion years of life's history. The human species is just one branch on an ancient evolutionary tree, and not so very different from some of the other branches—genetically we overlap more than 98 percent with chimpanzees. We are the product not of purpose, but of chance mutations. Bertrand Russell perfectly summed up man's place in the cosmos when he called him 'a curious accident in a backwater.'"[6]

One can see that what we have here is a fairly coherent and consistent story. It is the story of science as seen by the materialist. The main plotline is what may be called the "marginalization of man." Man starts with the religious view that he is in the center of things, and science rudely disillusions him.

There are at least two things wrong with this story: its beginning and its end. The beginning is not quite right. It is not really true that religious man saw himself at the center of the world, at least if we mean by religious man Jews and Christians. The idea of the world having a center entered Western thought not through Judaism or Christianity but through Greek astronomy and philosophy. The Greek picture, and more precisely the Ptolemaic-Aristotelian picture, was that the universe was arranged as a series of concentric spheres, with Earth at the center. However, the ancient Jewish picture of the world was vertical, not radial. God was above and the "abyss" was below. The human race was in between, both in place and in value. Among created things we were lower than the angels and higher than plants, animals, and inanimate objects. In no sense were we at the center. Indeed, having once been closer to God, human beings had been cast out. Man is an exile.

And even in the Greek picture of the cosmos, the central place was not the most exalted. On the contrary, the farther from the center things were the more sublime and beautiful they were, while the closer to the center the baser and grosser. More refined substances, like fire, tended upward, while heavy, earthy things tended toward their natural place at the center of the earth.

So the beginning of the story is really not right. Nevertheless, there is a kernel of truth in it. To the Jews and Christians human beings are certainly special among the visible creation, since we are "made in the image of God." However, this did not necessarily translate into a physical centrality.

What is more fundamentally wrong with the story, however, is its ending. If science had ended in the nineteenth century, the story would have some claim to accuracy. However, science did not end at that point. Instead, in the twentieth century it made discoveries even more profound and revolutionary than those of Copernicus and Newton. And, as a result, the story has become much more interesting. Like many good stories, the plot now has an unexpected twist at the end. It has not come out at all the way it was supposed to.

In fact, there has not been just one plot twist at the end, there have been at least five. In this book, I am going to tell these last and most interesting parts of the story. But it will be helpful for the reader to have some idea of the main story line before getting deeply involved in the various details and subplots, and so I will present here a brief outline or preview of the rest of the book.

FIVE PLOT TWISTS

The First Plot Twist

The first plot twist is similar to the standard man-no-longer-at-the-center-of-the-universe story in one respect: it has to do with our picture of the universe as a whole. However, it is not about space and whether it has a center, it is about time and whether it had a beginning.

We have already seen that it is at best an oversimplification to say that Jews and Christians placed man in the center of the cosmos while atheists and materialists placed man at the margins. But there is one contrast between the religious and materialist conceptions of the cosmos that has far more historical validity: Jews and Christians have always believed that the world, and time itself, had a beginning, whereas materialists and atheists have tended to imagine that the world has always existed. The very first words of the Bible, indeed, refer to a Beginning. By contrast, the pagan Greeks generally believed that the world was without beginning. (Among ancient thinkers, a few neo-Platonists believed that time had a beginning, but those who did were generally monotheists.)

Modern atheists and materialists have generally followed the pagan Greeks in this regard. The idea of a beginning of time was associated with religious conceptions, not with scientific theory, and those scientists who believed in a beginning did so for religious, not scientific reasons. Indeed, as we shall see, by the nineteenth century almost all the scientific facts seemed to point toward a universe without beginning. For this reason, the discovery of the Big Bang in the twentieth century came as a profound shock.

It is not that the Big Bang in itself proves the Jewish and Christian doctrine of Creation. Nevertheless, it was unquestionably a vindication of the religious view of the universe and a blow to the materialist view. It was as clear and as dramatic a beginning as one could have hoped to find. This is the story that is told in the next part of this book (chapters 4 through 8). The story is an ongoing one. Scientists are hoping to understand the Big Bang itself and to learn what, if anything, existed before it. I shall discuss some of these recent ideas as well.

The Second Plot Twist

In the standard materialist story of science, the world was found to be governed not by a personal God but by impersonal laws. Whereas religion had spoken of the earth and stars being fashioned by the hand of God, we now know that they were formed by blind and impersonal forces. Science looked not to God to explain events in the world but to physical "mechanisms" and processes. And this certainly remains true. It is still what the physical sciences are all about. But there has been a subtle and gradual shift in perspective that has become quite noticeable in the last century, especially in the more fundamental branches of physics.

Physics begins by trying to describe and explain various phenomena or effects observed in nature or in laboratory experiments. Empirical laws are discovered which these phenomena obey. As understanding progresses those empirical laws are found to follow from deeper laws and principles. For example, chemists in the eighteenth and nineteenth century discovered rules that govern the reactions of chemicals with each other, and formulated them in terms of such concepts as valency and chemical bond. In the early twentieth century those rules were discovered to be the consequence of the more fundamental laws of atomic physics. The laws of atomic physics, in turn, were found to flow from the laws of "quantum electrodynamics." And so it has proceeded as deeper and deeper levels of law have been uncovered.

As this deepening occurred, it had several consequences. First, physicists began to look not only at the physical effects and phenomena themselves, but increasingly at the form of the mathematical laws that underlie them. They began to notice that those laws exhibit a great deal of highly interesting mathematical structure, and that they are, in fact, extraordinarily beautiful and elegant from a mathematical point of view. As time went on, the search for new theories became guided not only by the detailed fitting of experimental facts, but also by these notions of mathematical beauty and elegance. A famous example is the discovery of the Dirac equation in 1928. The physicist Paul Dirac was seeking an equation to describe electrons in a way that would be consistent with the principles of relativity theory. In this search he was guided primarily by mathematical beauty. "A great deal of my work is just playing with equations and seeing what they give," he said. In this case, as he was playing with some equations he found something "pretty." "[It] was a pretty mathematical result. I was quite excited over it. It seemed that it must be of some importance." Notice that it was the "prettiness" of the mathematics that convinced him that he was on the right track. Soon after, he found the great equation that has been called "among the highest achievements of twentieth-century science."[7]

A second result of the deepening of physics has been the increasing unification of physics. Whereas in the early days of science the world seemed to involve many different phenomena with little apparent connection, such as heat, light, magnetism, and gravity, it later became increasingly clear that the laws of physics make up a single harmonious system. As physicists have gotten closer to finding the "unified theory" they have uncovered a great richness and profundity of mathematical structure in the laws of nature.

A consequence of these trends is that it is no longer just particular substances, or objects, or phenomena that physicists ask questions about, it is the universe itself considered as a whole, and the laws of physics considered as a whole. The questions are no longer only, "Why does this metal act this way?" or "Why does this gas act this way?" but "Why is the universe like this?", "Why are the laws of physics like this?" When it was just this or that effect or phenomenon, it was easy to say that physical laws were a sufficient explanation. It seemed out of place to talk about things being "fashioned by the hand of God." But when it is the laws of nature themselves that become the object of curiosity, laws that are seen to form an edifice of great harmony and beauty, the question of a cosmic designer seems no longer irrelevant but inescapable.

Psychologically speaking, atheism and materialism come from the natural tendency to "take things for granted," in particular to take the existence of the world itself for granted: the universe does not need to be created, it just is. It is taken to be a brute fact. But in the last century, physicists have learned to take less and less for granted. What could be more an unquestionable fact, for example, than the number of dimensions of space? And yet physicists now routinely study hypothetical universes that have only one or two dimensions of space, or that have dozens of dimensions. Indeed, the leading candidate for the ultimate unified theory says that there are in reality ten space dimensions. In the nineteenth century few people, if any, would have asked, "Why are there three space dimensions?" It was so much taken for granted that it was not even noticed as a fact that could be or ought to be explained. Today it is a central issue in physics.

Physicists even speak now about there being many universes, and suddenly the very number of universes is something one cannot take for granted either. It therefore becomes harder to avoid asking, "Why is there *any* universe at all?"

In the second main section of this book (chapters 9 through 13) I will look at the ancient Argument from Design for the existence of God. I will show how modern discoveries in physics, far from undermining that argument, have greatly strengthened it.

The Third Plot Twist

Perhaps nothing is so central to the materialist's story of science as the "dethroning of man." Not only did religion supposedly put human beings in the center of the universe, but it made us the very point or purpose of the universe. The universe was designed with us in mind. However, now, it is said, science has shown us that we were not "meant to be here." We were a fluke, our existence merely the result of "a fortuitous concourse of atoms." If science taught us anything, it seemed to be that. Except that, all of a sudden, science is telling us something very different.

In the 1970s, starting with some work of astrophysicist Brandon Carter, people started to talk about "anthropic coincidences." What they meant by this was that certain features of the laws of physics seem—just coincidently—to be exactly what is needed for the existence of life to be possible in our universe. The universe and its laws seem in some respects to be balanced on a knife-edge. A little deviation in one direction or the other in the way the world and its laws are put together, and we would not be here. As people have looked harder, the number of such "coincidences" found has grown.

Of course, this is exactly what one might expect if human beings *were* meant to be here, and if the universe was created with us in mind. And so, if nothing else, the discovery of these anthropic coincidences completely vitiates the materialist's claim that science has taught us otherwise.

This does not mean that the anthropic coincidences have ended the debate on the issue of man's place in the universe. In fact, there are ways that one can imagine explaining at least some of the anthropic coincidences without invoking the idea of a cosmic purpose. In other words, there is a "way out" for materialism. The way out goes by the name of the "Anthropic Principle," an idea that I shall explain later in the book. It is an idea that must be taken seriously. It is the explanation of the anthropic coincidences that seems to be preferred by most scientists who have written about them—and a considerable number of eminent scientists have. However, I shall argue later that the Anthropic Principle is itself problematic, and probably cannot completely explain away all the anthropic coincidences that have been found.

In any event, what is clear is that the moral that materialists have drawn from the story of science was prematurely jumped at. It looks very much like the story may turn out the other way.

The Fourth Plot Twist

If, as the materialist thinks, only matter exists, then the human mind must be a machine. The invention of the computer has made this idea appear more plausible to a great many people. Indeed, the intelligent robot or android has become a staple of popular science fiction entertainment. The defeat of the chess world champion Garry Kasparov by the computer program Deep Blue only confirmed for many the belief that it is merely a matter of time before computers become intelligent in every sense that human beings are. This seems to be the regnant orthodoxy in the field of artificial intelligence, or AI, and the professionals in that area have little patience with anyone who doubts it. Of course, if materialism is correct, and if the human mind is itself just a machine—a "wet computer" or "machine made of meat" as some have called it—then this confidence would have a great deal of justification.

However, there are many strong philosophical arguments against the notion that the human mind is no more than a physical machine. If it were, it would be very difficult to account for certain abilities of the human intellect, in particular the human abilities to think abstractly, to understand what philosophers call "universal terms," to know some truths with certainty, and to know of some truths that they are true "of necessity." We shall present these anti-materialist arguments later in the book. These arguments have an old pedigree. Some of them go back, at least in substance, to Aristotle, Augustine, and Descartes, although they have been developed further by many modern philosophers.

Unfortunately, scientific materialists seem to not pay much attention to such philosophical arguments. One reason for this is that many people who regard themselves as scientific tend to disparage the whole field of philosophy as futile and barren. They remember that physical science only began to make progress when it became divorced from metaphysics in the Renaissance.

But there has been a remarkable turn of events. In the twentieth century a powerful argument has been developed against the idea that the human mind is a computer, and this argument has come not from philosophy but from the science of computation itself. Specifically, the argument is based on a brilliant and revolutionary theorem proved in 1931 by the Austrian logician Kurt Gödel. Gödel himself regarded the idea that the human mind is a computer as merely a modern "prejudice." However, he did not develop the Gödelian argument against it in any detail, at least in print. That was done in the 1960s by the philosopher John Lucas, and in the 1980s and 1990s by the mathematician and physicist Roger Penrose. Both Gödel's Theorem, and the Lucas-Penrose argument that is based on it, are extremely subtle. But the gist of it can be summarized as follows. One consequence of Gödel's Theorem is that if one knew the program a computer uses, then one could in a certain precise sense outwit that program. If, therefore, human beings were computers, then we could in principle learn our own programs and thus be able to outwit *ourselves*; and this is not possible, at least not as we mean it here.

It's interesting to observe the lengths to which some people have been willing to go to avoid the conclusion of the Lucas-Penrose argument. Gödel's Theorem only applies to computers that reason consistently. The Lucas-Penrose argument can be defeated, therefore, by saying that the human intellect reasons in a way that is inherently *inconsistent*. Not just that human beings sometimes make lapses in logic, but that the human mind is radically and inherently unsound in its reasoning faculties. Many have been willing to say this. That anyone, to maintain a certain belief—in this case that he himself is a machine— would be willing to argue against his own mental soundness is startling. It seems to go beyond fideism and verge on fanaticism.

The Fifth Plot Twist

If the human mind is indeed a machine, and no more than that, it is clear that there can be no free will as that is normally understood. That is why most materialists simply deny that free will exists. And, in so denying, they once had what seemed an almost unanswerable argument from physics. The argument is that the laws of physics are "deterministic," in the sense that what happens at a later time is uniquely determined through the laws of physics by what happened at earlier times. If that is so, then all of human history, including every human thought and deed, could have been calculated (in principle) in advance from a knowledge of the way the universe was in the distant past—even before human beings appeared. And, for over two hundred years, from at least the time of Isaac Newton, every indication from physics was that the laws of physics *are* completely deterministic. Free will was, theoretically speaking, in a great deal of trouble. And that, in turn, created a painful difficulty for Judaism and Christianity, since human free will is a central tenet of those creeds.

However, a truly astonishing reversal came in the 1920s with the advent of quantum theory. Quantum theory was the greatest and most profound revolution in the history of physics. The whole structure of theoretical physics was radically transformed. And in that revolution physical determinism was swept away. The shock that this produced was immense. The ideal of physical science is prediction. Predictions are how theories are tested. To be able to explain the physical world is to be able to predict it in detail. Those ideas were deeply engrained in the minds of scientists. That complete knowledge of the present state of a physical system would *not*, even in principle, be enough to predict everything about its future behavior—which is what quantum theory showed—was a result that took the world of physics totally by surprise. The implications for the debate on free will were immediately and universally recognized. No longer could one simply argue from the deterministic character of physics that free will was an impossibility.

Of course, this has not ended the debate. Quantum theory certainly did not prove that there is free will. It simply showed that the most powerful argument against free will was obsolete. In the words of the great mathematician and physicist Hermann Weyl, "the old classical determinism . . . need not oppress us any longer."[8]

But this was only one of the remarkable reversals produced by the quantum revolution. In the opinion of many physicists—including such great figures in twentieth-century physics as Eugene Wigner and Rudolf Peierls—the fundamental principles of quantum theory are inconsistent with the materialist view of the human mind. Quantum theory, in its traditional, or "standard," or "orthodox" formulation, treats "observers" as being on a different plane from the physical systems that they observe. A careful analysis of the logical structure of quantum theory suggests that for quantum theory to make sense it has to posit the existence of observers who lie, at least in part, outside of the description provided by physics. This claim is controversial. There have been various attempts made to avoid this conclusion, either by radical reinterpretations of quantum theory (such as the so-called "many-worlds interpretation") or by changing quantum theory in some way. But the argument against materialism based on quantum theory is a strong one, and has certainly not been refuted. The line of argument is rather subtle. It is also not well-known, even among most practicing physicists. But, if it is correct, it would be the most important philosophical implication to come from any scientific discovery.

A New Story and a New Moral

G. K. Chesterton once compared his own intellectual development to the voyage of an English yachtsman "who slightly miscalculated his course and discovered England under the impression that it was a new island in the South Seas." The yachtsman of his story "landed (armed to the teeth and talking by signs) to plant the British flag on that barbaric temple which turned out to be the Pavilion at Brighton."

Those who manage to pass through intellectual adolescence all follow a journey that is somewhat like that. They are taught some simple truths as children, only to discover as teenagers or young adults that those truths were far too simple and that they themselves were embarrassingly simple to have accepted them. They strike off on their own, leaving the comfortable mental world of their childhood to find a wider and stranger world of ideas. They may experience this world as disturbing or as liberating, but in any event it is more exciting. If they are fortunate, however, they may come to rediscover for themselves the truths they were taught as children. They may return home, as T. S. Eliot put it, and know it for the first time. If so, they may see that, although they first learned these truths as simple children, neither the truths themselves nor the people who taught them were quite as simple as they supposed.

This requires, however, the difficult feat of questioning twice in one's life — of undergoing two revolutions in one's thinking. It requires being critical even of the ideas that one encountered in the first flush of critical thinking in one's youth.

The story of science has turned out to have involved such a double revolution. The revolutions of Copernicus and Newton have been followed by the revolutions of the twentieth century. Many have assumed that these further revolutions were just more of the same. Many have been misled by the name which

Einstein gave his theory into the absurd notion that his theory had something to do with "everything being relative." Many have been misled by the strangeness of modern physical ideas, such as quantum theory, into thinking that the lesson again was simply that all traditional notions should be jettisoned. However, a closer look at the scientific revolutions of the twentieth century reveals a very different picture. We find that the human mind is perhaps, after all, not just a machine. We find that the universe did perhaps, after all, have a beginning. We find that there is reason to believe, after all, that the world is the product of design, and that life is perhaps part of that design. This is the story that will be told in the rest of this book.

I should emphasize that this book is not about proofs. The materialist's story had a moral, but it did not constitute a proof of materialism. There was no experiment that proved that only matter existed, nor was there any calculation that proved that the universe had no purpose. Nor did the materialist really ever claim that there was. What he claimed was that there were two pictures of the world, the religious and the materialist, and that the progress of science has revealed a world that looks more and more like the materialist picture, and less and less like the religious picture. It was a question, in other words, not of proofs but of expectations. Science, it was claimed, had fulfilled the materialist's expectations and confounded the religious believer's. In this book I am making the same kind of claim, but in reverse. I am claiming that on the critical points recent discoveries have begun to confound the materialist's expectations and confirm those of the believer in God.

PART II

In the Beginning

4 The Expectations

One of the most dramatic discoveries in the history of science is that the universe began in an explosion that took place about 15 billion years ago. The evidence for this "Big Bang Theory" has become so strong in the last thirty years that it is no longer seriously questioned. Now, if it is true that the Big Bang was the beginning of time itself, as at least appears to be the case, then one of the central beliefs of Jews and Christians has been confirmed, and one of the assumptions that had prevailed among scientific materialists has been overthrown.

Where did these beliefs and assumptions come from? For Jews and Christians, belief in a beginning comes from the very first words of the Bible: "In the beginning, God created the heavens and the earth." There are really two ideas contained in this statement. The first is that the universe had a beginning in time, or in other words that it has a finite age. The second idea is that the universe was created by God. Most people tend to think that the first idea is contained in the second: if the universe was created, then surely it must have had a beginning too. There is a lot to be said for this way of thinking. Certainly, most things in our experience seem to confirm it. For example, if we know that some object was made by a human artisan, then we also know that it had a beginning.

However, this link between creation and beginning is not quite so clear when it comes to the creation of the universe by God as taught by Judaism and Christianity. In saying that God is the Creator of the universe, what is being said is that he is the cause of the universe, that he is the explanation of the fact that the universe exists. Now, if one thinks about it for a while, one can see that a thing can be caused without necessarily having had any beginning in time. For example, imagine that an object is illuminated by a lamp. The lamp is the cause or explanation of the object's being illuminated. However, nothing in that fact tells us whether the lamp has been illuminating the object for a finite time or for infinite time. If the lamp has always been illuminating the object, then the illumination of the object had no beginning, but nevertheless it always had a cause.

It is not clear, then, from the assumption that God causes the universe to exist that anything can be proven one way or the other about whether the universe had a beginning. Philosophers have long argued about this. St. Thomas Aquinas, for one, believed that while the existence of a creator or "First Cause" could be proven philosophically, the universe having a beginning in time could not be. In his *Summa Contra Gentiles*, for example, he has a chapter entitled "On the arguments by which some try to show that the world is not eternal."[1] There he examines six such arguments, only to reject them all as "lacking absolute and necessary conclusiveness." His view was that, while "it was entirely fitting that God should have given created things a beginning," it was hypothetically possible for God to have created a universe that had no beginning in time. (For a longer discussion of the Jewish and Christian notion of Creation see appendix A.)

The real reason, then, that Jews and Christians believe that the universe had a beginning in time, is that they think that this has been explicitly revealed by God in the first words of Genesis: "In the beginning, . . ." Of course, as we have seen, many things in the Book of Genesis have been interpreted allegorically, even in ancient times. However, the assertion that there was a "beginning" is not one of those things. Jews and Christians have been virtually unanimous in understanding these words as teaching a literal, temporal beginning. In the year 1215, the Fourth Lateran Council declared that it was a matter of Christian faith that the world was created by God *"ab initio temporis,"* i.e., "from the beginning of time."[2]

Not surprisingly, religious believers greeted the discovery of the Big Bang with satisfaction. Most scientists of the materialist stripe, on the other hand, were made acutely uncomfortable by the discovery, as we shall see later from some of their comments. Most materialists had expected that the world would turn out to be eternal, in the sense of having no beginning and no end.[3] Why so? There are probably several reasons. One of them is essentially psychological. If something has always been around, then it is easy to take it for granted and not question its existence. But if something makes a sudden appearance, we naturally ask ourselves why.

When my oldest son, Thomas, was two years old, one evening I pointed out to him the bright yellow orb of the moon hanging low in the sky. The moon was a novelty for him, and he promptly asked me, "Who put it up there?" Of course, adults know that astronomical bodies were formed by gas and dust condensing gravitationally. But if I had told my son that, he could have raised the question of who put the dust, the gravity, and the space there. He could have asked, in short, "Why is there a universe at all?" Seventy years ago, had he and I been around back then and had he asked that question, I might have been able to silence him with the reply, "The universe has always been here." What made the Big Bang Theory unpleasant for some people, I suspect, is that it took away that answer.

This brings us to a more philosophical reason for the discomfort many materialists have with the idea that time had a beginning. Scientific materialists tend to think of the explanation of things entirely in terms of finding the physical processes or mechanisms that produced those things. Of course, this is a perfectly good mode of explanation, and it has served very well in the physical sciences. As noted in chapter 3, the Scientific Revolution was made possible by the fact that physicists and astronomers abandoned more philosophical modes of explanation in favor of "mechanistic" ones. Before that, scientists had gone in for "teleology," that is, explaining the behavior of things in terms of the goals or ends that they were seeking, or the purposes they served. This works pretty well for people and animals, but proved to be an unproductive way to think about inanimate objects.[4] Physical objects are not seeking a certain kind of future, rather they are driven by forces and events in the past and present.

The materialist therefore tends to feel that the explanatory job is done when he can give an account of the physical processes by which the present state of affairs arose out of some past state of affairs. As long as he can trace the chain of physical causation farther and farther back in time, he is satisfied. If the universe is infinitely old, then each physical event that ever happened had a past out of which it grew and in terms of which it can be explained. Of course, such a scheme of explanation is not really complete. The reason that it is not is well stated by astrophysicist P. C. W. Davies in his book *God and the New Physics*:

> Suppose horses had always existed. The existence of each horse would be causally explained by the existence of its parents. But we have not explained yet why there are horses at all—why there are horses rather than no horses, or rather than unicorns, for example. Although we may be able to find a [past] cause for every event . . . , still we would be left with the mystery of why the universe has the nature it does, or why there is a universe at all.[5]

The point being made by Davies is somewhat subtle, however, and is easily overlooked. It is all too simple for one who has narrowed his intellectual vision and who is interested only in chains of physical causation to be misled into thinking that materialism affords a complete scheme of explanation of reality. He can maintain this illusion as long as he is able to keep his nose down on the trail of past physical causes. This is why the idea of an eternal universe is comforting to him: if time had no beginning then the trail he is doggedly pursuing will never run out, he can follow it ever deeper into the past and thus forestall indefinitely the point at which he must face the more fundamental questions of why the universe has the nature it does and why it exists at all.

In any event, the historical fact is that Jews and Christians believed in a beginning of time, while scientific materialists strongly preferred the idea of an ageless universe.

5 How Things Looked One Hundred Years Ago

As scientific discoveries accumulated in the centuries leading up to our own, the expectations of the materialist seemed to be confirmed. There was no evidence that the universe had a beginning, and several discoveries seemed to indicate that it had always existed.

To begin with, in Newtonian physics, time was conceived of as a single dimension stretching without limit into both past and future, just as space was conceived of as an infinite three-dimensional volume stretching without limit in every direction. In Newtonian physics, it was natural to assume that time goes on forever just as numbers go on forever, from $-\infty$ to $+\infty$.

Later, "the principle of conservation of energy"—the First Law of Thermodynamics—was discovered. As we all learned it in school, "energy can never be created or destroyed, but only changed in form." I vividly remember that when I was ten and had just learned this principle in school, I used it to argue against an older brother, that the universe did not have to be created because the energy in it could always have existed. Indeed, I argued, the law of conservation of energy says that this energy *could not* have been created. I suppose this was a clever argument for a ten-year-old. I certainly thought so at the time.

In the nineteenth century, chemists discovered the law of conservation of mass. In chemical reactions, the total mass of the reactants does not change—at least as far as could be measured back then. Therefore, not only energy but matter itself was both indestructible and uncreatable. Eventually, with the equivalence of mass and energy, which was discovered by Einstein, these two principles were subsumed into a single principle. No violation of this principle has ever been observed.

To all appearances, therefore, both the world of space and time, and all the matter and energy in it, had always existed and always would. To say that time had a beginning, while not absolutely ruled out, would have seemed very strange from the viewpoint of nineteenth-century physics. It would have been as strange as saying that space did not go on forever but had edges to it somewhere. At that time, there was no hint of the "beginning" in which religious people believed. The materialists' expectations seemed to have been borne out. As we shall see, however, the picture changed in quite a dramatic and unexpected way as a result of Einstein's Theory of General Relativity.

6 The Big Bang

The first hint of the Big Bang—but a hint not understood at the time—was a discovery by an American astronomer with the improbable name Vesto Melvin Slipher. In 1914 he announced to a meeting of the American Astronomical Society in Evanston, Illinois, a remarkable finding. All of the galaxies that he had studied up to that time, about a dozen, were receding from the earth at fantastic speeds—some up to 2 million miles an hour. By 1925 he had extended his survey to include forty-five galaxies, and almost all showed the same strange behavior.

About this time, another American astronomer came into the picture: Edwin Powell Hubble. With Milton Humason, a mule-train driver and janitor turned astronomer, he used the powerful one-hundred-inch-diameter telescope at Mount Wilson Observatory in California to measure the movements of the galaxies. Not only did Hubble and Humason confirm Slipher's findings, they went beyond him in a very important way: in addition to measuring the velocities of galaxies, they were able to estimate how far away they were. This is not an easy thing to do. One cannot stretch out a measuring stick to another galaxy. But, in a clever indirect way, Hubble and Humason were able to infer these distances, and thus they discovered two remarkable things.

The first thing was that the galaxies were much farther away than anyone had imagined. Some were millions of light-years away. (One light-year is the distance that light, which travels at 186,000 miles per second, traverses in a year. A light-year is about 6,000,000,000,000 miles.) This discovery can be seen as part of the story told in chapter 3 of the marginalization of man. This story has an interesting twist. Hubble and Humason actually, for technical reasons, misjudged cosmic distances by about a factor of ten. They therefore thought the other galaxies were closer, and smaller, than they truly are. As a result, our own galaxy—

the Milky Way—seemed to be considerably larger than all the rest, a miscalculation which maintained some degree of anthropocentrism. When the mistake was later remedied, however, it was found that the Milky Way is a fairly typical-sized galaxy after all.

The second remarkable thing that Hubble and Humason found was that the farther away a galaxy is from us the faster it is receding from us. In fact, the distances and recession speeds are roughly proportional to each other—this is called Hubble's Law. Hubble's Law may seem to put Earth in a special location, but that is an illusion of perspective. One can show that if Hubble's Law is satisfied by observations made at one place in the universe, like Earth, it will be satisfied by observations made at any other place. A way to see this is to imagine the galaxies painted on a rubber sheet, and then to imagine that the sheet is stretched uniformly to twice its size. Two "galaxies" that were an inch apart will recede another inch, while two that were a foot apart will recede another foot. That is, the farther apart they start, the faster they recede—Hubble's Law. And it is apparent that this will be true for any pairs of galaxies on the rubber sheet; none are singled out as special.

Hubble's Law has an important implication. If one traces the movements of the galaxies backward, one finds that at some point in the past all the galaxies were on top of each other. That is, the Hubble expansion suggests that all the matter in the universe is flying apart from some primeval explosion.

Hubble and Humason's discoveries were announced in 1929. The true significance of them is far more profound than appears at first sight, and can only be understood in the light of Einstein's theory of gravity, the so-called General Theory of Relativity. So let us now turn from observation to theory, and from 1929 back to 1916.

What Einstein had proposed in that year was that space and time, instead of being merely some static backdrop against which events played out, were themselves something dynamic and changing. Whereas Newtonian space and time were like a rigid screen on which things moved around, Einsteinian space-time was a flexible fabric, which could bend and quiver in response to the matter and energy that moved around on it. Einstein wrote down a set of equations that describe exactly how space-time bends and quivers:

$$G^{\mu\nu} = 8\pi G_N \, T^{\mu\nu}$$

This looks like one equation, but actually contains several equations packaged within it. Therefore, physicists often talk about "Einstein's equations." The quantity $G^{\mu\nu}$ is called the "Einstein tensor" and describes the curvature or flexing of space-time, while the quantity $T^{\mu\nu}$ is called the "stress-energy tensor" and describes how matter and energy are distributed in space and time. G_N is Newton's constant, a number characterizing the strength of the force of gravity, which also appeared in Newton's law of gravitation. To understand Einstein's equations one has to know a great deal of physics and mathematics, but just looking at them one can see that there is something rather simple about them. Even so, apples falling from trees, planets orbiting the sun, the vast whirlpool motion of galaxies, which take hundreds of millions of years to go around once, the behavior of "black holes"—objects so strange that time and space get mixed up inside them, even the expansion of the universe itself, are all described by this single, simple-looking set of equations.

It is the last application of Einstein's equations that concerns us now. What Einstein realized is that not only could the fabric of space-time bend and stretch in one locality, but that the whole of space, the entire universe indeed, could stretch like an expanding balloon. In this balloon analogy, one is not supposed to be thinking of the air or space *inside* the balloon (or outside the balloon, for that matter). It is the elastic sheet of the balloon itself that is supposed to represent the space of our universe, and physical objects are—as we spoke of the galaxies before—like little spots of paint on the balloon's surface. Of course, the balloon is only a two-dimensional sheet, while the space of our universe, which it is supposed to represent, is three-dimensional. But do not worry, the mathematics of differential geometry, which is used to study curvature, can be applied to any number of dimensions. For the purposes of analogy, two dimensions are quite adequate. (One might ask whether there is anything about the real universe that corresponds to the air or volume inside—or outside—the balloon. That is, is our three-dimensional universe embedded in some higher-dimensional space, just as the two-dimensional sheet of the balloon is contained in the air around it? No, not at least in the standard cosmological theory. It is not easy, but one is supposed to imagine just the surface of the balloon *without* imagining the inside and outside of it.)

When a balloon is expanded, the area of the rubber sheet increases—there is more balloon surface. In the same way, the universe can expand so that there is more universe. The actual volume of the universe can increase (or, for that matter, decrease). At this point, I am describing what is called a "closed universe," which has a finite volume at any given time. It is possible for the universe to have a finite volume, just as the balloon has a finite surface area; and the universe can curve around on itself, just as the surface of the balloon curves around to form a finite area that has no edges. In a closed universe, if one were to extend a line in any direction, it would eventually come back around near to where it began, in the same way that a line drawn on the rubber sheet of the balloon will come around back to where it started. On the other hand, there are also so-called "open universe" solutions to Einstein's equations, in which the universe has an infinite volume at any given time. An open universe would be like a rubber sheet that does not curve back on itself, but extends to infinity in all directions. These "open universes" will be important in some of our later discussions.

A very important point to realize is that in Einstein's theory an expansion of the universe does not simply involve an expansion of the matter, in the sense of the matter sliding through space away from other matter. It is the space separating the matter that is expanding. In the balloon analogy, it is not that the painted spots move along the rubber surface of the balloon, but that the balloon itself is stretching to put more space between the spots. If one traced the motion back one would not come to a time when all the galaxies were crowding on top of each other surrounded by empty space.[1] Rather, one would come to a point where all the *space* was on top of itself, so to speak, where the volume of space was zero. That is, the whole balloon started off being of *zero* size.[2]

The amazing fact, then, is that the time at which the expansion began—if such a time existed—was also the time at which space and time themselves began. In an expanding universe described by Einstein's equations, time and space themselves had a beginning. But we are getting ahead of the history here. We are still in the year 1916.

After he discovered his equations, Einstein quickly realized that they cannot, in the form he originally wrote them down, describe an eternal static universe. In Einstein's equations, the shape and movement of space is described by the Einstein tensor, $G^{\mu\nu}$. What the matter—galaxies and the rest—is doing is described by the stress-energy tensor, $T^{\mu\nu}$. In essence, through Einstein's equations the matter tells the space what to do. The important point is that if the universe is filled with matter, as it is, space cannot remain static. It must be expanding or contracting. A static universe filled with matter is impossible, because all the matter pulls on all the other matter gravitationally.

There is an analogy with a ball that is up in the air. If there is no gravitational pull from the earth, a ball could remain suspended at a fixed height above the ground. But because of gravity, you know that if you see a snapshot of a ball up in the air, it is either on its way up or on its way down. It can be motionless, at most, for an instant, if it happens to be at the top of its motion.

The realization that the universe must be dynamic profoundly disturbed Einstein. Like most materialists, he had naturally assumed that the universe was static and eternal. Some may object to calling Einstein a materialist, since he often spoke of "God." However, the word *God* as he used it did not signify a personal being. He believed, he said, in the same kind of God as the philosopher Spinoza, an immanent, somewhat abstract deity. Perhaps one can come close to Einstein's ideas by saying that "God" for him stood for the orderliness and harmony of the universe. (Nevertheless, he often spoke in very personal terms of his God, referring to him as "The Old One." A famous statement of Einstein's was, "Subtle is the Lord, but he is not malicious."[3]) In the final analysis, for

Einstein the universe with its beauty and harmony is really all there is; it encompasses all of reality, and therefore he conceived of it as being eternal.

In order to avoid the strange conclusion that the universe was expanding or contracting, Einstein modified his equations in 1917. He added a new term to them so that they looked like this:

$$G^{\mu\nu} = 8\pi G_N T^{\mu\nu} + 8\pi G_N \Lambda\, g^{\mu\nu}$$

The new term is the one containing the parameter Λ. The curious thing about this term is that it has no effect whatsoever on ordinary gravitational phenomena. It has, for example, no effect on the way apples fall, or planets orbit, or galaxies spiral around. In fact, it has one and only one physical consequence — it affects the way the universe *as a whole* expands or contracts. For this reason Λ is called the "cosmological constant." The cosmological constant (if it is positive) gives a repulsion that tends to counteract the mutual gravitational attraction of the matter. If the cosmological constant has precisely the right value, then the new equations do admit solutions in which the universe is static and eternal, with an exact counterbalancing of the effects of matter and the new cosmological term.

The addition of this cosmological constant term was a perfectly logical step for Einstein to take. In the first place, the original mathematical and physical reasoning that led Einstein to his equations allowed for the possibility of such a term. There was no theoretical reason, therefore, not to include it. Moreover, at that time there was no evidence that the universe was expanding or contracting. Nevertheless, Einstein himself came to regard his addition of the cosmological constant term to his equations as a major mistake. In fact, he called it the greatest blunder of his life. He felt that he had needlessly complicated his equations and somewhat marred their beauty. Had he instead stuck with the original form of his equations he could have *predicted* the expansion of the universe and the Big Bang.

Once the expansion of the universe was discovered by astronomers the cosmological constant fell by the wayside. It was no longer needed, and physicists generally came to believe that it was not there. By a curious twist of fate, however, evidence has come from astronomical observations in just the last few years that suggest strongly that there might be a cosmological constant after all. And its value seems to be very roughly what Einstein proposed. One could say then that Einstein's "greatest blunder" has turned out instead to be a masterstroke. And yet, even though there may well be a cosmological constant, it does not fulfill the purpose for which Einstein introduced it. It does not exactly counterbalance the attraction of ordinary matter to produce a static universe. In fact, in 1930 Arthur Eddington showed that such an exactly balanced static universe would not be stable.[4] So while Einstein turned out in the end not to be wrong to introduce a new term into his equation, he was wrong in thinking that the universe was static and eternal.

The first people really to take seriously the idea of an expanding universe were Alexander Friedmann, a Russian meteorologist and mathematician, and Georges Lemaître, a Belgian physicist and Catholic priest. Friedmann, in 1922, and Lemaître, independently in 1927, discovered the solutions to Einstein's equations that described an expanding universe filled with matter.[5] At first, Einstein did not believe Friedmann's result, and published a proof that it was wrong. Shortly afterward, however, he admitted that his proof was in error, and that Friedmann was correct.

It was Lemaître who tied it all together by relating his expanding-universe solutions of Einstein's equations to the ongoing observations of Hubble and Humason at Mount Wilson Observatory. Lemaître also proposed that all the matter in the universe was originally concentrated into an incredibly dense "primeval atom" that exploded to produce the world we see. He thus laid down the basic outlines of our present "hot Big Bang" theory.

Only in 1930 was Einstein forced to admit that what he had so long resisted was probably true—fourteen years after he proposed his theory of gravity, and eight years after Friedmann's work. He said, "New observations by Hubble and Humason concerning the red shifts of distant nebulae [i.e., galaxies] make it appear likely that the general structure of the Universe is not static."[6]

Einstein's initial attitude to an expanding universe had been expressed in letters to Willem de Sitter, a Dutch physicist: "One cannot take such possibilities seriously." Such feelings were shared by many others, even after the discoveries of Hubble and Humason. Eddington, in 1931, wrote, "The notion of a beginning is repugnant to me. . . . I simply do not believe that the present order of things started off with a bang . . . the expanding Universe is preposterous . . . incredible . . . it leaves me cold." The German physicist Walter Nernst wrote, "To deny the infinite duration of time would be to betray the very foundations of science."[7]

In the words of the cosmologist Andrei Linde, "The nonstationary character of the Big Bang theory [based on the] Friedmann cosmological models seemed very unpleasant to many scientists in the 1950s."[8] As late as 1959, a survey of leading American astronomers and physicists found that two-thirds of them believed that the universe had no beginning.[9] This was forty-three years after Einstein's equations were discovered, and thirty years after Hubble and Humason's results.

There can be no question that the aversion that some scientists felt for the Big Bang Theory stemmed largely from philosophical prejudices, and in particular to the fact that the reality of a beginning seemed to sit much better with religious views than with their own materialism.[10]

While reluctance to accept the Big Bang Theory was largely a matter of philosophical prejudice, it was not entirely that. The theory faced, for a while, a significant problem. As mentioned, Hubble had gotten the distance scales wrong by a factor of ten. The galaxies were therefore thought to be ten times closer to each other than they really are. As a result of this mistake, the age of the universe was miscalculated by the same factor of ten: it was estimated to be not about 15 billion years old, as now believed, but closer to 2 billion years old. This was a problem, because there were also estimates of how old the earth was and how old stars were that were several times greater than this. Obviously, the universe itself cannot be younger than the objects it contains.

Partly in response to this problem, Hermann Bondi, Thomas Gold, and Fred Hoyle proposed the so-called Steady State Theory in 1948. According to their hypothesis, the universe had existed for an infinite time and had always been expanding just as it is now. In fact, their theory said that it had always looked essentially the same as it does now, which is why it was called the Steady State Theory. This sounds like a contradiction, for if the universe is expanding, the matter in it should get ever more thinned out as the galaxies get farther apart from each other. What Bondi, Gold, and Hoyle proposed was that matter was continuously being created to make up for the thinning out due to the cosmic expansion. The universe, therefore, while always expanding, could always remain of the same density.

The Steady State Theory sounds rather bizarre in retrospect. Nevertheless, it was taken seriously by many scientists for reasons that were as much philosophical as narrowly scientific. Aside from solving the problem that the age of the universe appeared too small in the Big Bang Theory, the Steady State Theory was consistent with a strong form of something called, rather grandly, the Cosmological Principle. The Cosmological Principle said that every part of the universe was pretty much the same as every other part of the universe. This was an extrapolation from the trend of discoveries since Copernicus. Earth is not at the center of the universe, as Ptolemy thought, but is in a place that is completely undistinguished. The Sun is an ordinary star, in an ordinary galaxy. If we went to any other part of the universe, we would find similar galaxies, similar stars, and similar planets.

In its weaker version, the Cosmological Principle only held that *at any given time* every part of the universe is similar to every other part. This is also the assumption made in the standard Big Bang cosmological model. In the jargon of the field, the universe is assumed to be "homogeneous and isotropic." The stronger version of the Cosmological Principle, however, said not only that there is nothing special about *where* we live, but that there is nothing special about *when* we live. The uniformity of the universe extends through time as well as space. In other words, the idea is that things have always been essentially the same as they are now, and always will be. Of course, Earth itself had a beginning and will have an end. But there have always been planets like Earth and stars like the Sun, and there always will be.

This Strong Cosmological Principle was not something that came out of scientific observation or scientific theory. At root it is just a philosophical prejudice — in fact, it is really nothing but the prejudice in favor of an eternal universe. This is not to say that those scientists who found it appealing were being bigoted or unscientific. When the evidence against the Steady State Theory got too strong that theory was abandoned, more quickly by some people, but eventually even by its founders. However, while it lasted, the theory made some very strong demands on the credulity of its adherents. The continuous creation of matter that it entailed was a very radical proposal, because it meant abandoning the principle of conservation of matter and energy, which to physicists is something almost sacred. Indeed, it is somewhat ironic that a principle that had once seemed to imply an eternal universe now had to be abandoned, in the face of all evidence, to maintain belief in an eternal universe. Some of the other proposals to avoid a Big Bang, such as the "tired light theory," involved equally radical modifications to the laws of physics.

THE BIG BANG CONFIRMED

In 1948, two students, Ralph Alpher and Robert Herman, working with George Gamow, started thinking about what the universe must have been like in its earliest moments if there had been a Big Bang. They realized that it must have been intensely, inconceivably hot. The volume of the universe was much smaller right after the Big Bang than it is today, and therefore matter was in a state of extreme compression, leading to tremendous heat. One second after the Big Bang, for example, the density of matter was several thousand times the density of lead, and the temperature was about 10 billion degrees centigrade. The early universe was a raging inferno.

The heat of this "primeval fireball" meant that the universe was filled with intense radiation. Indeed, in technical jargon, the universe was "radiation dominated" for the first several million years of its existence. Much of this radiation was in the form of light. There is some historical irony here. Once, it was a common argument against the literal interpretations of Genesis that light was created on the "first Day," while the sun and stars were not made until later, on the "fourth Day." It now appears that the biblical chronology was quite right in this respect. Light indeed existed virtually from the beginning,[11] while stars took many millions of years to appear.

As the universe expanded, this primordial radiation cooled and became more faint. The afterglow of the Big Bang was "red-shifted" to longer and longer wavelengths. Now, 15 billion years after the Big Bang, it is no longer in the form of "visible light" but in the form of microwaves (the kind of light that heats your food in microwave ovens). The universe is filled with this ghostly remnant of the flash with which it began. Shortly after World War II, Gamow, Alpher, and Herman predicted that such a "cosmic background radiation," as it is now called, should exist, but no one paid much attention until the mid-1960s, when a group at Princeton led by Robert Dicke had the idea of trying to detect it. However, as it turned out, someone else beat them to the punch.

In 1965, radio astronomers Arno Penzias and Robert Wilson, working at Bell Labs in New Jersey, made a series of measurements with a detector that had been specially designed for satellite communications. They found a noise, or static, that seemed to come equally from all directions in the sky. At first they thought that it was a problem with the device itself, or some local interference — they even considered the possibility that heat given off by bird droppings inside the antenna was responsible. Eventually, however, because of the work going on in nearby Princeton, the true significance of what they were seeing was realized. They were hearing a whisper from the Big Bang. Since that time, several pieces of strong confirming evidence have been found, and the fact of the "hot Big Bang" is no longer much disputed.[12]

7 Was the Big Bang Really the Beginning?

Most physicists tend to think of the Big Bang as really being the beginning of the physical universe, and with it, the beginning of time itself. This idea of a "beginning of time" is hard to grasp. The mind wants to ask, "What happened before the Big Bang?" In the standard Big Bang model that question is meaningless: there was no time before the Big Bang, and thus no such thing as "before" the Big Bang. Perhaps an analogy will help. What lies to the north of the North Pole? Nothing. The highest latitude that it makes sense to talk about is 90 degrees. There is no such place as 91 degrees latitude. Latitude runs out at the North Pole. Let us call the moment of the Big Bang $t = 0$. Then there are no negative times; the time coordinate simply runs out at $t = 0$. The point is that time is not some absolute thing that exists whether there is a universe or not, just as latitude does not exist unless there is a globe. Time, like space, is a measure of the intervals between things and events in the physical universe, just as latitude is a measure of intervals on the earth's surface. This, at least, is what time means to a physicist. One cannot talk about the time between events if there is no universe and therefore are no events.

As hard as this idea is to grasp, it was grasped 1,600 years ago by St. Augustine. In a celebrated chapter of his *Confessions*, he reflected deeply upon the nature of time. The beginning of his reflections was the question, "What happened before God created the heavens and the earth?" This question was posed as a trap by pagan opponents of Christianity and Judaism. Why, they asked, did God wait around an infinite length of time before creating the world?[1] St. Augustine began by pointing out that time itself is merely an aspect of the created world: "There can be no time without creation."[2] Therefore, he argued, time itself, no less than the universe, is something created: "What times could there be that are not made by you, [O God]?"[3] Remarkably, he even grasped that there was no time before the universe began: "You [O Lord] made that very time, and no time could pass by before you made those times. But if there was no time before heaven and earth, why do they ask what you did 'then'? There was no 'then,' where there was no time."[4] This great thinker grasped a point that was not fully appreciated until the coming of twentieth-century developments in physics. Many of St. Augustine's statements about time have a strangely modern sound and seem pregnant with meaning to scientists who think about relativity. Indeed, the *Confessions* is sometimes quoted in technical papers on quantum cosmology,[5] and Bertrand Russell praised St. Augustine for what he called his "admirable relativistic theory of time."[6] (It is relativistic in a certain sense, but not in the full technical sense of the theory of relativity.)

It should be emphasized that this modern discovery of a beginning of time was a vindication for Jewish and Christian thought.[7] While religious thinkers, like Augustine, spoke of a beginning to time, and while this concept was enshrined in Christian doctrine, no one until the twentieth century could make scientific sense out of such an idea. How could time have a beginning? It not only seemed absurd, but was, quite literally, unimaginable. Unimaginable, and yet, at least in the standard cosmology of today, true.

Even though the Big Bang, in standard cosmological theory, is the beginning of time, it is important to realize that alternative, very speculative cosmological scenarios have been proposed in which things did happen before the Big Bang. I will briefly explain the basic ideas behind three of these scenarios: the bouncing universe, baby universes, and eternal inflation. It is not essential for the discussions that will come later to read these sections, and so some readers may wish to skip directly to the next chapter. However, the ideas explained in the rest of this chapter have some interest of their own, and are not without relevance to some points made later in the book. Before explaining the basic ideas of such speculative scenarios as the bouncing universe, baby universes, and eternal inflation, I have to explain some of the possibilities that exist within the conventional or standard cosmological model. In particular, in the standard cosmological model does the universe have finite size or infinite size? Will it stop expanding at some point, or will it expand forever? These are the questions I will deal with in the next section.

THE UNIVERSE IN THE STANDARD BIG BANG MODEL

Homogeneity and Isotropy

In most theoretical discussions of cosmology the universe is assumed to be at least approximately "homogeneous and isotropic." "Homogeneous" means that the universe looks the same from any place in it. More precisely, if one considers any particular time in the history of the universe, then *at that time* physical conditions, such as the temperature, the density of matter, or the curvature of space, will be the same at all points of space. "Isotropic" means that the universe looks the same in every direction.

There are two reasons why cosmologists often assume the universe to be homogeneous and isotropic. The first is that it greatly simplifies matters for them. If at a given time the density of matter is the same everywhere in the universe, it can be specified by just one number, "the matter density of the universe." The same thing is true for other physical quantities, such as the curvature of space, or the temperature. The second reason that they make this assumption is that the universe is in fact quite homogeneous and isotropic on large scales. It is true that on what cosmologists consider "small" scales, such as the scale of the solar system or even that of the Milky Way, the universe is not at all homogeneous and isotropic. Conditions inside the Sun, for example, are very different from those on Earth. And conditions within the solar system are very different from those in interstellar space. However, if one considers distances much, much greater than the size of galaxies, the universe does indeed appear quite uniform. Once one averages over such small details as stars, planets, galaxies, and clusters of galaxies, the universe does appear reasonably homogeneous and isotropic. For the rest of the discussion in this section, I will be considering only idealized models of the universe in which it is assumed to be *perfectly* homogeneous and isotropic.

The Curvature of Space

The first thing we will consider is the curvature of space. If we are looking at a two-dimensional space, or surface, we know intuitively what it means for it to be curved. For example, the surface of a sphere is curved. By contrast, a plane is flat. But what does it mean to say that a three-dimensional volume of space is "curved" or "flat"? Here we cannot appeal directly to intuition, because human beings are not equipped to visualize curved spaces of more than two dimensions. But we can get some idea of what is meant in the following way.

If we draw two lines that are going in the same direction on a flat plane — i.e., "parallel lines" — they will never meet. They will always remain the same distance apart. If, however, we draw two lines *on a sphere* that start off going in the same direction, they will not remain the same distance apart, they will draw closer and closer together. (For example, if two lines on a globe start at the equator and go due north, they begin by going in the same direction, but as they are extended northward they get closer together until they intersect at the North Pole.) Are there surfaces where two lines that begin by going in the same direction end up getting *farther* apart? Yes. Consider, for example, the end of a trumpet, where it flares out. If two lines were drawn on the trumpet so that they began by going in the same direction, the flaring of the trumpet would cause these lines to splay out from each other and diverge. This behavior of "parallel" lines can be used to define what we mean by curvature. When parallel lines always stay a fixed distance apart, we say that the surface has zero curvature, i.e., it is flat. When they always end up converging, as on a sphere, we say that the surface has positive curvature. And when they always end up diverging, as on a trumpet-shaped surface, we say that the surface has negative curvature. The more quickly the lines converge or diverge, the greater the curvature is said to be.

This can all be made very precise mathematically. In fact, this is how mathematicians think about curvature; and the beauty of it is that the same ideas can be used to explain what we mean by three-dimensional spaces being curved. If a three-dimensional space is such that when two lines start off going in the same direction they always end up moving closer together, we say that the space is positively curved; if they always end up moving farther apart, we say that it is negatively curved; and if they always remain a fixed distance apart, we say that the space is "flat."

Now, a two-dimensional surface can certainly be curved more in one place than another—like the surface of the earth, which has flat places and hills and valleys. Or it can have exactly the same curvature everywhere, like a perfect sphere. The same possibilities hold for spaces of higher dimensionality. In particular, the three-dimensional space of our universe could vary in curvature from place to place. However, if the universe were perfectly homogeneous and isotropic, as we are idealizing it to be, it would have to have exactly the same curvature everywhere. There are then three possibilities: (1) it has the same positive curvature everywhere (analogous to a sphere), (2) it has zero curvature everywhere (like a flat plane), or (3) it has the same negative curvature everywhere.

Open and Closed Universes

Just as a sphere has a finite surface area, so too does a universe whose space has the same positive curvature everywhere have a finite volume of space. Such a universe is called spatially "closed." But a "flat universe" with zero curvature everywhere and a universe with the same negative curvature everywhere each have an infinite volume of space and are spatially "open."

Is our actual universe open or closed? No one knows. As of now, the data are consistent with its being "flat." (Remember, we are talking about its behavior on extremely large scales. Near lumps of matter, like Earth, or the Sun, or a galaxy, matter will warp space. But on very large scales the universe appears to be quite flat.) The data are still quite crude, and do not allow us to tell whether the universe really has zero spatial curvature on the whole, or has instead a slight positive or negative curvature. We may never know whether the universe has a truly infinite or a vast but finite volume.

Many people who have heard that the universe is expanding think that this implies that the universe has a finite volume right now. After all, they reason, if the universe were already of infinite volume, it would make no sense to talk of its getting any bigger. Reasonable as this sounds, it is wrong. The universe can be infinite in size and yet be expanding. To understand this, let us think about two-dimensional analogies.

The surface of a sphere has a finite area. If the sphere expands, as the surface of a balloon does when it is inflated, then its surface area increases. Now consider an infinite flat plane. We can imagine that this plane is an infinite piece of graph paper and has grid lines drawn on it. We can further imagine that this infinite piece of graph paper is put into an infinitely large photocopying machine that is set to a magnification of 120 percent. The copy made by the machine will also be an infinitely large piece of graph paper, but the grid lines will all be increased in spacing by 120 percent. Any picture drawn on the original paper will also be scaled up by 120 percent. In a real and simple sense, we can say that the copy is 120 percent "larger" than the original, even though both have infinite area. Or, to be closer to the balloon analogy, we can imagine an infinite rubber sheet with grid lines or other figures drawn on it. The sheet can be stretched uniformly in every direction, so that the grid spacings and figures get ever larger, even though the sheet always has infinite area.

When cosmologists say that the universe is expanding they in no way mean to imply that it is finite in volume. The expansion makes objects that are very distant from each other grow farther apart—the whole universe scales up as if by a cosmic copying machine that is making ever larger copies. But we have no idea at present whether the universe is infinite or finite in extent—i.e., whether it is open or closed.

Whether the universe is open or closed is closely connected to how much mass or energy (Einstein tells us the two are equivalent) it contains. If the density of mass/energy in the universe has a certain "critical value," then the space of the universe is flat, and therefore open. (We are assuming here as always that the universe is homogeneous and isotropic.) If the mass/energy density of the universe is larger than the critical value, then space has a positive curvature and is closed. If the mass/energy density is less than the critical value, the universe is open. What matters here is the total mass/energy, which includes both mass/energy in what we shall call "ordinary matter"—i.e., electrons, protons, atoms, particles of light, and so on—and the mass/energy associated with the "cosmological constant" Λ that we spoke of previously.

Will the Universe Expand Forever?

We know that the universe is presently expanding. Will that expansion last forever or will it at some point stop and even reverse? The future of the universe depends on how much mass/energy it has and what kind of mass/energy it is filled with. Ordinary matter tends to slow down the expansion of the universe. One can understand this in a simple intuitive way. Ordinary matter tends to attract other ordinary matter gravitationally, and so the mutual attraction of the ordinary matter in the universe acts to pull the universe together, i.e., resist its expansion. If the universe were filled with only ordinary matter its expansion would always be slowing down. In fact, if its density is more than the critical value we mentioned, then the expansion will eventually stop and reverse. The universe will then begin an irresistible collapse that will culminate in a "Big Crunch" that is the reverse of the Big Bang. Space and time itself will come to an end. So far, I have been discussing the case where all mass/energy in the universe is in the form of "ordinary" matter (and the universe is homogeneous and isotropic). In that case things are relatively simple, and there is a connection between whether the universe is open or closed (i.e., infinite in spatial volume or finite) and whether it will expand forever. If it is infinite in volume it will expand forever, whereas if it is finite in volume it will eventually collapse.

Things are very different and more complicated if (as seems to be the case) some of the mass/energy in the universe is in the form of "cosmological constant." Then there are other possibilities. For example, one could have a universe that is closed (finite in volume) but expands forever. The reason that things are more complicated is that the cosmological constant, if it is positive, tends to cause a repulsion that makes the universe expand even faster, as mentioned before. In fact, this seems to be actually happening. Observations suggest that earlier in its history the expansion of the universe was slowing down, but that "recently," in cosmologist's terms, i.e., in the last several billion years, it has picked up again and is accelerating. The best evidence that we now have suggests the following things about our universe: (1) it is "flat" or nearly so, but we don't know yet whether it has a tiny positive spatial curvature or a tiny negative one; therefore, (2) it may be infinite in volume or finite, we cannot yet tell which; and (3) its expansion is accelerating, suggesting that it will expand forever. (However, one can imagine possibilities in which the acceleration will stop and the universe eventually collapses.)

THE BOUNCING UNIVERSE SCENARIO

One can think of the expansion of the universe as following a trajectory, like a ball shot out of a cannon. (The cannon is the Big Bang.) If the universe is filled only with ordinary matter its mutual gravitational pull tends to slow the expansion down, just as the earth's mass pulling on the cannonball tends to slow its ascent. If the density of matter in the universe is above some critical value, then the universe will eventually fall back upon itself. Similarly, in the cannonball analogy, if the mass of the earth is great enough (in relation to the cannonball's height and speed of ascent), the cannonball will fall back down upon the earth. On the other hand, if the earth's pull is not great enough, the cannonball will keep going up forever—it has "escape velocity." In the same way, if the density of matter in the universe is subcritical, the universe will expand forever. This is all assuming that the universe is filled with only ordinary matter. A positive cosmological constant acts in a different way from ordinary matter, as we saw. It tends to accelerate the expansion of the universe. In the cannonball analogy, it would be as if the cannonball were jet-assisted.

The ball analogy very naturally suggests the following idea. If a ball falls back to earth it can bounce; so perhaps if the universe falls back upon itself in the Big Crunch it too can bounce. In other words, at the moment of the Big Crunch, matter—and space itself—might rebound and start expanding again. This expansion would look just like a new Big Bang. And eventually, this new expansion would also reach a maximum extent and the universe would begin to collapse again to another Big Crunch, followed by another rebound, and so on, perhaps forever.

This suggests the further idea that there may already have been an infinite number of such cycles of expansion and contraction in the past. If that was so, then the universe is "eternal" after all, and the Big Bang, rather than being the beginning of time, was only the commencement of the latest cycle. This idea is not at all unreasonable, and it is an interesting possibility, to which, quite naturally, some theoretical attention has been given. However, it faces several difficulties.

In the first place, given the laws of physics as we know them, it is not clear that the universe could bounce. Certainly, it would not if Einstein's theory of gravity in its classical (i.e., non-quantum) form were used to calculate what happens at a Big Crunch. In 1969, Stephen Hawking and Roger Penrose proved mathematically that space-time at a Big Crunch becomes, in the mathematical jargon, "singular." Specifically, the curvature of space-time becomes infinite, and "world-lines" come to an end. If this is true, then bounces do not happen. However, there may be ways that such a "singularity" might be avoided by nature. In particular, extremely close to the time of a Big Crunch—or Big Bang—(in fact, within 10^{-43} seconds of it) "quantum effects" are believed to be important. These may very well "soften" the singularity in some way—we will return to that idea later. But whether they would allow an actual bounce is not known, and in fact this will remain unanswerable until the correct theory that unifies Einstein's theory of gravity with quantum theory is known.

Even if the universe does bounce, there is a problem with the idea that these bounces have extended back infinitely into the past. The Second Law of Thermodynamics suggests that the "entropy," or the amount of disorder, of the universe will be greater with each successive bounce. The universe would not be born completely fresh in each bounce, so to speak, but would show its age. Given this, it would seem quite unlikely that the universe has already undergone an infinite number of bounces in the past.

Finally, present data indicate an expansion of the universe that is getting faster and faster, and therefore suggest that the expansion will go on forever rather than reversing and leading to a crunch or bounce. In any event, beyond all of these objections, which are very cogent, the bouncing universe idea simply has not helped physicists to solve any theoretical problems, and consequently it no longer receives a great deal of attention. (However, there has been a very interesting recent attempt to revive it.[8])

So far, all three attempts to avoid a beginning that I have discussed—Einstein's carefully adjusted cosmological constant, the Steady State Theory, and the bouncing universe scenario—were clearly motivated to a large extent by philosophical prejudice against the idea of a beginning; and all of them are either discredited now or strongly disfavored. However, more recent ideas like baby universes and eternal inflation are still possibilities, and at least some of them have better scientific motivation.

The Baby Universes Scenario

The baby universes idea is basically simple in concept: the universe splits. In the balloon analogy, one can picture a little part of the balloon being pinched off and forming another separate balloon. That little baby balloon could then eventually expand to enormous size. In this picture, the Big Bang could have been the event where our universe was pinched off from a larger one. One can envision a situation where this process of universes giving birth to other universes has been going on forever.

The Eternal Inflation Scenario

What Is Cosmological "Inflation"?

The first thing to do is explain the idea of "inflation" itself. This will also help us to understand a few points that will be discussed in later chapters. Inflation is one of the most important and beautiful ideas in modern cosmology. It was proposed in 1979 by Alan Guth to resolve a number of serious cosmological conundrums. These included the so-called "horizon problem" and "flatness problem."

The horizon problem refers to the fact that the "cosmic background radiation" discovered by Penzias and Wilson is extremely uniform. It glows with almost exactly the same intensity in every part of the sky. To understand why this is puzzling, we must start by recalling what this radiation is. It is the afterglow of the primeval fireball. Most of this glow was given off when the universe was about three hundred thousand years old. After that point the fireball cooled and became relatively transparent. The cosmic background radiation that we now see is therefore reflecting conditions as they were early in the history of the universe. What it tells us is that the region of the universe that we are now able to observe was very homogeneous back then. More precisely, the density of matter, the temperature, and the pressure were uniform to a few parts in a hundred thousand.

At first this does not seem too strange. After all, if you are sitting in a room, the air in that room is also very uniform in density, temperature, and pressure. But the reason for that is that any pockets of higher pressure or temperature would quickly smooth themselves out by the flow of air and heat from one part of the room to another. Air will flow from regions of high pressure to regions of lower pressure, and heat will flow from hotter to colder. But such flows take time, and in the early universe there was not enough time for flows of heat and matter to produce the uniformity we now observe. The word *horizon* refers to how far one can see. The point is that when the universe was only three hundred thousand years old, any regions that were more than a three hundred thousand light-years apart could not "see" each other, because there had not been enough time for light to travel from one of those regions to the other. Not only could such regions not see each other, they could not influence each other in any way. They were, in cosmology jargon, outside each other's "horizon." And yet, they were somehow at almost exactly the same temperature, pressure, and density. This is the "horizon problem."

The flatness problem has to do with the fact that space itself is much less curved than it has any right to be in the standard model of cosmology. As we have explained, "flat" here does not mean two-dimensional, it means lacking in curvature. Einstein tells us that matter, such as planets, stars, and galaxies, warps the space-time in its vicinity. However, if the universe is looked at on much larger scales of distance its space appears to have very little, if any, curvature. This fact long puzzled cosmologists.

The hypothesis of inflation resolves both the flatness problem and the horizon problem. The idea of inflation is that *very* early in its history—within a tiny fraction of a second after the Big Bang—the universe, or a least some large part of it, underwent a brief and extremely rapid period of expansion. Not the gradual expansion that is seen by astronomers today, but a sudden jump in size by an enormous factor in a tiny fraction of a second. This "inflationary phase" is imagined to have lasted only a very brief time, after which the more normal, gradual type of expansion that we see today is supposed to have resumed.

The way inflation solves the flatness problem is easy to understand. If we were to blow up a balloon to enormous size, its surface would become very flat, just as, because of its enormous size, Earth's surface was for a long time thought to be flat. In the same way, if just after the Big Bang the universe was suddenly hugely expanded, space would exhibit very little curvature thereafter.

Inflation solves the horizon problem in a similar way. The universe may well have had a very non-uniform appearance shortly after the Big Bang. There may have been great variation in physical conditions such as temperature, pressure, and density from one place to another. But if one looked at a small-enough patch of space, conditions would have been uniform in that patch. What inflation is thought to have done is take one tiny little uniform patch of space and stretch it to such enormous size that it entirely contains the part of the universe that we can now see with the most powerful telescopes—the part, that is, which is within our horizon. That is why things look so uniform within our horizon.

This aspect of inflation is relevant to the discussions about "anthropic coin-cidences" in a later chapter. The important lesson that inflationary theories have taught us is that the sameness that characterizes the part of the universe that we can see may be a misleading indicator of the way things are throughout the *whole* universe. It is quite possible that the universe looks very different—not just in temperature or pressure, but in much more radical ways—in regions that are too far away for us to observe, even though things look remarkably uniform in the part we can observe.

This idea of the whole universe, or a very large part of it, suddenly increasing in size by a huge factor in a fraction of a second sounds bizarre, to say the least. Nevertheless, there are some very reasonable ideas for how this may have hap-pened, based on well-established ideas in physics. For example, shortly after the Big Bang, the universe was filled with matter at high temperature and density. This matter could have undergone what is called a "phase transition." Common examples of phase transitions from everyday life are the melting of ice and the boiling of water. Certain kinds of phase transitions in the early universe would have caused the "stress-energy" of matter, which appears in Einstein's equations, to suddenly change in such a way as to cause space to inflate.

So, however bizarre it may seem, inflation is a very reasonable hypothesis from the viewpoint of modern physics. Two words of caution are nevertheless in order. First, many different inflationary scenarios have been proposed. Just to name a few broad categories, there have been "old inflation," "new inflation," "supersymmetric inflation," "Kaluza-Klein inflation," "extended inflation," "chaotic inflation," "open inflation," and "hybrid inflation." Some of these variants of inflation have been shown not to lead to a realistic picture of the universe, but many of them are still viable. Moreover, within each of these categories, many specific models have been invented. There is no standard theory of inflation. Second, while recent observations have provided very strong indirect evidence for inflation, the evidence is not yet such as to leave no room for doubt. But the evidence in favor of it is getting stronger all the time.

How Inflation Could Happen Eternally

The idea of "eternal inflation" was proposed by the physicist Andrei Linde in 1986. The basic idea can be explained using the by-now-familiar balloon analogy. Think of the universe as the surface of a giant balloon that is gradually expanding. One can imagine that small patches on this balloon are undergoing rapid "inflationary" expansion. If they do, they will grow to form huge blisters, as it were, on the fabric of space. In fact, these blisters can inflate to such a size that they are really themselves like large balloons. After a while, these blisters can stop their rapid inflation and begin to expand in a normal, gradual manner. However, small patches on those blisters may continue to inflate rapidly. In this way, each blister will develop huge inflating blisters on *it*. Eventually, *these* will stop inflating and expand normally, but when they do small patches of space on *them* may continue to inflate, forming blisters on blisters on blisters. This process can go on and on indefinitely. And, conceivably, it may have been going on forever. However, there are physical arguments that suggest that it cannot have been.[9]

Each such blister on the space of the universe can eventually grow to enormous size. The region of the universe that we inhabit and that we can observe with telescopes may be just a part of one of these patches that inflated. What we call the Big Bang may just have been the process by which a new blister formed and inflated to astronomical proportions.

8 What If the Big Bang Was Not the Beginning?

We have seen that there are a variety of scenarios in which what we call the Big Bang was only the start of one part or phase of a more encompassing reality, but not the beginning of everything, not the beginning of time itself.

Will we ever be able to tell whether the Big Bang really was the beginning of time? Almost certainly not by direct observation. Even if something existed before the Big Bang, the extreme conditions that prevailed in the first moments after the Big Bang almost certainly effaced any record of it. We cannot peek behind the veil. If there were other bounces before the Big Bang, we will not be able to tell directly. If there are new "universes" forming as blisters on our universe, as in the eternal inflation scenario, they are certainly much too far away for us to see. They would lie outside of our horizon. If our universe was pinched off from another universe altogether, as in the baby universes scenario, there is no way to see the universe from which ours sprang, since it was disconnected from ours by the pinching off.

In fact, it is quite typical of such scenarios with many universes or domains that the entities they postulate *cannot* be directly seen by us. This is not something that should cause anyone to mock these ideas. They are very serious ideas, and one of them may someday be shown to be right. However, it should make religious believers less embarrassed by the fact that some of the entities they talk about also cannot be dragged into the laboratory.

Nevertheless, suppose that it could be shown somehow, indirectly, that the Big Bang was not the beginning of time. That would still not resolve the issue of whether there *was* a beginning of time; it would only show that the Big Bang itself was not that beginning. The Big Bang may have been just a bounce, but that does not tell us whether there was a first bounce. The Big Bang may have been the time when our universe started forming as a blister on a pre-existing space, but that does not tell us whether that pre-existing space itself had a beginning. In fact, it is hard to imagine how it could ever be determined with certainty whether time itself had a beginning or not.

One can hope that someday we may find the ultimate laws of physics. It may even be possible to solve their equations and see whether they imply bounces, blisters, or baby universes. But that will not necessarily answer the question about the beginning. For example, even if the laws of physics allow the universe to bounce, the bouncing could have started with a first bounce, or it could have been going on eternally. Whether time had a beginning may be forever out of reach of an empirical or theoretical resolution.

However, this book is not about what we will know someday; it is about what we know now, and what trends we can discern in discoveries up to this point. And one trend is clear: everything that we have ever studied has proven to have had a beginning.

Individual human beings have beginnings, of course. That is obvious from personal observation, for we see children being born. However, it could be imagined that the communities into which individual human beings are born have always existed. But, of course, this turns out not to be the case. Historical research is able to show that villages, tribes, and nations too have beginnings.

Again, it is conceivable that the civilizations of which these villages, tribes, and nations form a part are themselves eternal. But this also can be refuted by historical investigation. Every civilization had a beginning. Is it possible, at least, that there has been an endless series of civilizations? Hypothetically, yes, but archeology shows that there was a time, about five thousand years ago, before which no civilizations existed.

Nor has the human race been around forever, as Aristotle thought. *Homo sapiens* appeared a few hundred thousand years ago. No skeletons of *Homo sapiens* older than that have been unearthed. And while *Homo sapiens* is but one branch on a much more ancient tree of life, that tree, too, had a beginning about 4 billion years ago. Even the very matter from which life emerged had a beginning: there was a time before which there were no atoms, or even protons, neutrons, and electrons. They were forged in the furnace of the Big Bang.

Astronomers know that stars are born and die. As with human beings, there are different generations of stars. New stars form out of the remains of stars that have died. Yet the generations of stars also had a beginning. Until recently, it appeared as though the universe, at least, was ageless. This too has now been proven false. Will it turn out that the "universe" that began with the Big Bang is actually part of some larger universe? Quite possibly. Will that larger universe prove to be something eternal? The entire history of discovery would lead one to doubt it.

There is another fact that would lead one to doubt the world's eternity. This fact is a very fundamental one about our physical universe called the Second Law of Thermodynamics. What this law basically says is that physical processes tend to run down. Early in any physical process, energy is in a more organized, "usable" form. As time goes on this energy gets degraded into a more disorganized, unusable form. In technical language, "entropy" increases. A simple example is that a rolling ball will slow down, and eventually stop, as its energy of motion is worn down by friction and converted into heat.

One never sees the reverse happening, however. A ball rolling on a level surface will not roll faster and faster, and by some kind of "reverse friction" convert heat back into rolling motion. The Second Law of Thermodynamics is the reason why those processes which have a direction only go in *one* direction. Metal rusts, it does not unrust. Milk spills, it does not unspill. China cups break, they do not unbreak. In general, things age; they wear out; they decay. Rocks erode, paper grows yellow and crumbles, clothing becomes threadbare, memories fade. Nothing withstands "the ravages of time." That is why physicists believe that, in our universe at least, a "perpetual motion machine" is an impossibility. And it is a powerful reason to doubt that the universe is eternal. If it is, then it itself is a kind of perpetual motion machine.

What science has shown us is that not only the things we make, but even the things of nature which seem immortal and unchanging are subject to this universal aging: even the Sun and stars and the very fabric of space itself.[1] This would not have surprised the author of Psalms:

> Of old hast thou laid the foundation of the earth; and the heavens are the work of thy hands.
>
> They shall perish, but thou shalt endure: yea, all of them shall wax old like a garment. (Ps. 102:25–26)

The facts that science has taught us give strong reason to doubt that the universe is eternal. Admittedly, a rigorous proof in this matter may never be possible, but the question we are asking in this book concerns expectations rather than proofs. The claim which we are disputing is that the progress of knowledge has revealed a world that ever more conforms to the expectations of materialists, and ever less to those of religious believers. The claim is that this has been the trend of science. And, indeed, one must honestly confess that the trend in the physical sciences did appear, as of a century or two ago, to favor the idea of an eternal universe. However, one would have to say now that the trend of discovery points very much the other way. As Cardinal Ratzinger has accurately written,

> [The theory of entropy,] the theory of relativity, and . . . other discoveries . . . showed that the universe was, so to speak, marked by temporality—a tem-

porality that speaks to us of a beginning and an end, and of the passage from a beginning to an end. Even if time were virtually immeasurable, there would still be discernable through the obscurity of billions of years, in the awareness of the temporality of being, that moment to which the Bible refers as the beginning—that beginning which points to him who had the power to produce being and to say: 'Let there be . . . ,' and it was so.[2]

PART III

Is the Universe Designed?

9 The Argument from Design

Our discussion now shifts to another aspect of our universe, not where it came from but what it is like. I have already said that God as "First Cause" is not only a historical cause, the cause of the beginning (if any) of the universe, but a continuing cause, the cause that each and every instant of the universe *is*. And just as the universe around us may show traces of creation in the first sense, so it may show traces of creation in the second sense. That is, just as its existence may point to the God who gave it being, so its structure may point to the God who designed it.

THE COSMIC DESIGN

The human authors of the Bible were not scientists. They were not interested in the relationships among natural phenomena. For them, the supremely important relationship was the world's dependence upon God, its creator. The world was God's handiwork, and everything in it pointed to him and was a reflection of his wisdom and power. The biblical authors saw God's hand in everything, not only in what we would call miracles, but in quite ordinary events of nature such as rain or snow or earthquakes, which they portrayed as being the result of direct divine interventions. Moreover, in their descriptions of the world, they used the cosmological ideas familiar to them and to their contemporary readers. These ideas, which were those of an ancient, pre-scientific people, naturally seem very primitive to us.

And yet, as non-scientific as the Bible was in its outlook, in a number of ways its message helped to clear the ground and prepare the soil for the much later emergence of science. It did this in part by overthrowing the ideas of pagan religion. In paganism, the world itself was imbued with supernatural powers and populated with capricious beings: the Fates and Furies, dryades and naiades, sun gods and gods of war, goddesses of sex and fertility. All of these were swept away by the severe monotheism of the Bible.[1]

This explains a historical fact which sounds rather strange at first: the pagan Romans accused the early Christians of the crime of "atheism" or "godlessness," and persecuted them for it. The reason for this was that Judaism and Christianity *were* — in the small g sense — "godless." All of the things which the pagan had learned to venerate as divine were reduced to the status of mere things by Jewish and Christian teaching. The Sun was not a god, but merely, according to the Book of Genesis, a lamp. The animals and other good things of Earth were not to be worshipped by human beings, but rather human beings were to exercise dominion over them. Whatever reverence, awe, or wonder that it was appropriate to have for the ocean or the stars or living things was not on account of any divinity or spirituality that they possessed, but because they were the masterworks of God.

It is often said that science "dis-enchanted" the natural world, in the sense of depersonalizing it and desacralizing it. But to a large extent this had already been accomplished by the Hebrew Bible. The universe was no longer alive with gods, but was a work of cosmic engineering.

As a work of engineering the cosmos necessarily reflected the rationality and wisdom of its creator. This is a constant biblical theme. According to numerous passages in the Bible, it was the divine Wisdom that presided over the creation of the heavens and the earth. In the Book of Proverbs, Wisdom sings of her own role in creation:

> From everlasting I was firmly set, from the beginning, before earth came into being.
> The deep was not, when I was born, there were no springs to gush with water.
> Before the mountains were settled, before the hills, I came to birth;
> Before he made the earth, the countryside, or the first grains of the world's dust. . . .
> I was by his side, a master craftsman, delighting him day after day, ever at play in his presence,
> At play everywhere in his world. (Prov. 8:22–26,30–31)

Another biblical theme of central importance is that God is a lawgiver. Indeed, Jews call the first five books of the Bible "the Torah," meaning "the Law." It was this law that regulated the Covenant between God and the people of Israel. But the Torah was understood by the ancient rabbis to be much more than a set of books written on parchment or rules written to guide human conduct. The Torah was a law that existed before the world began and was the master plan according to which God created the universe.[2] In the words of the rabbis, "the Holy One, blessed be he, consulted the Torah when he created the world."[3] The universe, then, was conceived of and created by God according to a law. This law, indeed, was identified by the rabbis with the divine Wisdom.[4]

This conception of God as the lawgiver not only for human beings but for the cosmos itself is found explicitly in various passages of the Bible. Through the prophet Jeremiah the Lord says: "When I have no covenant with day and night, and have given no laws to heaven and earth, then too will I reject the descendants of Jacob and of my servant David." (Jer. 33:25–26)

According to Psalm 148, the Sun, the Moon, the stars, and the heavens obey a divinely given "law, which will not pass away." Moreover, there is even a hint that this law is mathematical in nature. St. Augustine and, later, the writers of the Middle Ages laid great stress on a passage from the Book of Wisdom which says that God has "disposed everything according to measure, and number, and weight." (Wis. 11:20)

The church's supposed hostility to science has become such a standard part of anti-religious mythology that many will be taken aback by the idea that Jewish and Christian concepts played a role in preparing the way for science. Yet that is what the work of such scholars as Pierre Duhem, A. C. Crombie, and Stanley Jaki has helped to show. It is becoming more generally realized that it was not an accident that the Scientific Revolution occurred in Europe rather than in the other great centers of civilization. For example, in his recent book *Consilience*, the biologist E. O. Wilson, discussing the fact that the Chinese civilization, with all its refinement and splendid achievements, did not produce a Newton or a Descartes, had this to say:

> Of probably even greater importance, Chinese scholars abandoned the idea of a supreme being with personal and creative properties. No rational Author of Nature existed in their universe; consequently the objects they meticulously described did not follow universal principles, . . . In the absence of a compelling need for the notion of general laws—thoughts in the mind of God, so to speak—little or no search was made for them.[5]

This idea of God as cosmic lawgiver was from very early times central to Jewish and Christian thinking.[6] It is the basis of the so-called Argument from Design for the existence of God. An early statement of this argument can be found, for example, in the works of the Latin Christian writer Minucius Felix near the beginning of the third century:

> If upon entering some home you saw that everything there was well-tended, neat and decorative, you would believe that some master was in charge of

it, and that he was himself much superior to those good things. So too in the home of this world, when you see providence, order and law in the heavens and on earth, believe that there is a Lord and Author of the universe, more beautiful than the stars themselves and the various parts of the whole world."[7]

It is the beauty, and order, and law that we see in the world which point to its creator.

Over the centuries there have been many other statements of the Argument from Design by believers of every confession. The idea that the universe is a work of supreme craftsmanship or engineering is expressed, for example, in numerous passages of Calvin's *Institutes* such as this: "[W]hithersoever you turn your eyes, there is not an atom of the world in which you cannot behold some brilliant sparks at least of his glory. . . . You cannot at one view take a survey of this most ample and beautiful machine in all its vast extent, without being completely overwhelmed with its infinite splendour."[8] A classic statement of the Argument from Design was given by the Anglican theologian William Paley (1743–1805) in his book *Natural Theology*. He compared the world to a watch, whose intricate structure proved that it did not arise by accident:

> In crossing a heath, suppose I pitched my foot against a *stone*, and were asked how the stone came to be there: I might possibly answer, that, for anything I knew to the contrary, it had lain there forever; nor would it, perhaps, be very easy to show the absurdity of this answer. But suppose I found a *watch* upon the ground, and it should be inquired how the watch happened to be in that place. I should hardly think of the answer I had given before—that, for anything I knew, the watch might always have been there. Yet why should not this answer serve for the watch as well as for the stone?[9]

Paley went on to observe that the universe possesses a fineness of structure far surpassing that of any watch or other human contrivance. Therefore, like a watch, it must have been designed.

In 1840, a few decades after Paley, the great English historian and essayist Thomas Babington Macaulay wrote:

> A philosopher of the present day . . . has before him the same evidences of design in the structure of the universe which the early Greeks had . . . for the discoveries of modern astronomers and anatomists have really added nothing to the force of that argument which a reflective mind finds in every beast, bird, insect, fish, leaf, flower, and shell.[10]

Two Kinds of Design

In modern times the Argument from Design for the existence of God has come in for a great deal of criticism. Much of this criticism is made by people who claim that the designs seen in nature have a purely natural, scientific explanation. Before we get into that debate — in the next chapters — it is important to make some distinctions.

Comparing the various statements quoted in the previous section, one can see that there are really two versions of the Design Argument. One of them, which I shall call the Cosmic Design Argument, is based on the order exhibited by the cosmos as a whole: Jeremiah spoke of a law given to "heaven and earth," and Minucius Felix of the providence, order, and law "in the heavens and on earth." The most common examples given in the older texts seem to be from astronomy: the law obeyed by the Sun, Moon, and stars, the "covenant with day and night," and the beauty of "the stars themselves." Macaulay, too, began by referring to "the structure of the universe," and mentioned the discoveries of modern astronomers. And yet, all of the specific examples that Macaulay listed are taken from biology. All the "structures" are those of the bodies or parts of bodies of living things: "beast, bird, insect, fish, leaf, flower, and shell." In these examples, Macaulay was making what I shall call the Biological Design Argument. It is interesting that the Biological Design Argument seems to have come much later historically than the Cosmic Design Argument.

Closely related to this distinction between the cosmic and the biological versions of the Design Argument is a basic ambiguity in the meaning of the word *structure*. There are really two quite different kinds of structure. On the one hand, there is the kind that comes to mind when we hear the words *regularity*, *pattern*, *symmetry*, and *order*. It is describable by mathematical rules. This is the kind of structure one sees, for example, in decorative patterns, in the steps of a dance, or in the form of a sonnet. In the natural world, one can see it in the shapes of crystals and in the regular motions of the solar system, to give just two instances. This is what I shall call "symmetric structure."

But there is another kind of structure, which we find in living things, and which I will therefore call "organic structure." *Webster's* dictionary defines an "organism" this way: "a complex structure of interdependent and subordinate elements whose relations and properties are largely determined by their function in the whole."[11] Note that organisms are inherently *complex*. Even a bacterium, the simplest kind of living thing, has, according to molecular biologists, an enormously complicated structure. By some reckonings, it would take as great a quantity of information as there is in the *Encyclopedia Britannica* to describe all the parts of a bacterium and how they are put together. And multicelled creatures, of course, are vastly more complicated still.

A second feature of organic structure is that the various parts are not related to each other by some mathematical rule. There are formulas that tell you how the molecules in a crystal are arranged, or even the rhyme scheme of a sonnet. However, there is no formula that tells you how liver and lungs and eyelash and tendon are related to each other in the body of an animal. Rather, as *Webster's* says, the parts of an organism are related to each other by their respective *functions* in a complex and interdependent whole.

Artificial machines also exhibit organic structure. An automobile, for instance, has a great variety of parts: steering wheel, headlights, windshield, tires, carburetor, spark plug, camshaft, and so on. As in a living thing, these are related to each other by function.

Both in living organisms and artificial machines, it is not just the parts but the whole itself which has a function or purpose. The function of a car is to get people around. The function of a mosquito, according to evolutionary biologists, is to compete for resources and make more mosquitoes.

It should be emphasized that the two kinds of structure are not mutually exclusive. For example, suspension bridges and buildings have complex functional designs, but they also exhibit simple mathematical patterns. The same is true of living things. One need only think of the spiral shape of the nautilus's shell, the fivefold symmetry of the starfish, the hexagonal pattern of an insect's compound eye, or the helical structure of the DNA molecule.

The universe in which we live has many examples of both kinds of structure, the symmetric and the organic, often combined in the same object, as in the "leaf, flower, and shell" mentioned by Macaulay; and both kinds of structure have been seen as evidence of a cosmic designer. It is the first kind that Calvin doubtless had in mind when he wrote that "the exact symmetry of the universe is a mirror, in which we may contemplate the otherwise invisible God,"[12] and the second kind when he wrote that "the composition of the human body is universally acknowledged to be so ingenious, as to render its Maker the object of deserved admiration."[13]

This is why the comparison chosen by William Paley of a timepiece is so apt. The regular circular motions of a watch's parts remind one of that cosmic clock we call the solar system, while the intricacy of a watch's structure and the complex functional interdependence of its parts resemble the structure of a living thing.

10 The Attack on the Argument from Design

The Design Argument for the existence of God has been attacked on the ground that structure can arise in a purely natural and spontaneous way without any planning or any intervention by an intelligent being. In fact, at least three different ways are known by which structure can arise naturally: (1) pure chance, (2) the laws of nature, and (3) natural selection.

PURE CHANCE

Let us return to William Paley wandering about his heath picking up stones. If the stones scattered about the heath are randomly shaped, then it is quite unlikely that on a brief walk Paley will find any interesting ones. However, if he examines a great number of stones he has a chance of finding one that is peculiarly shaped. He may find one that is very symmetric, or one that looks like it has some purpose. I imagine that most of us have come across a stone that bore an accidental resemblance to a primitive tool.

In other words, given enough chances complicated patterns are bound to appear accidentally. This is the old "monkey with a typewriter" idea. According to probability theory, if a monkey were to type randomly for an unlimited time, eventually it would accidentally type the play *Hamlet*, or indeed any other text. How long would it be before that monkey typed the play *Hamlet*? The answer is that the number of pages the monkey would have to type to have a chance of typing *Hamlet* is so huge, that just to write down that number would require a book longer than the play *Hamlet* itself. Even if we were only to wait for the monkey to accidentally type a particular fairly short sentence it would take an inconceivably long time. For example, in order for the monkey to have a reasonable chance of accidentally typing the words *Mary had a little lamb*, it would have to type about 1,000,000,000,000,000,000,000,000,000 pages.

In view of these large numbers, it seems implausible that any structure of significant complexity would arise entirely by chance. However, I shall return to this point later, because despite the seemingly impossible odds there is a way that it might happen.

THE LAWS OF NATURE

Now, if Paley were to come across a stone that was cubical or rhombohedral in shape it would not be too surprising. Many crystals have those shapes, and there is a perfectly good explanation of how crystals form naturally. Under certain conditions, atoms or molecules do arrange themselves, without any help, into very regular arrays. This happens when water freezes, for example. When the temperature of water is above the freezing point, the water molecules move around in a random way. But when the temperature is lowered to the freezing point, they begin to line up rank upon rank in a hexagonal pattern. This happens automatically. Aimless molecular motions end up producing a highly ordered structure. There is no traffic cop directing this process; it is a natural consequence of the laws of physics.

Another example is the regular motion of the planets around the Sun. The orbits of most of the planets are very close to being circular. Moreover, these circles all lie almost in the same plane, so that the solar system looks like a giant pinwheel or platter. As we shall see later, this orderly arrangement and pattern of motion emerged in a natural way out of the chaotically swirling cloud of dust and gas which cooled and condensed to form the planets. Again, the orderliness of the final arrangement did not require any intervention in the process by an intelligent being; it required only the operation of the laws of physics.

NATURAL SELECTION

The kinds of processes just mentioned, where the laws of physics lead in a fairly direct way to the formation of structure, produce only the kind that I called symmetric structure. The freezing of liquids and the gravitational condensation of dust clouds lead to elegant patterns but they do not lead to complex organisms. For a long time there was really no good idea to explain how organic structure could arise in a natural manner. Probably this was one reason why Macaulay, in the passage quoted in chapter 9, laid particular stress on biological structures, like flowers, leaves, and shells, as being evidence of a designer. However, very shortly after Macaulay wrote that passage Charles Darwin published his monumental work *On the Origin of Species.*

The significance of Darwin's work was not simply that it challenged the scientific accuracy of the accounts given in the Book of Genesis. That is a relatively minor thing. We have already seen that even in ancient times there was a great deal of variation in how those accounts were interpreted, and many writers of high authority in the church read them in a very non-literal way. The deeper challenge posed by Darwin's theory was the idea of "natural selection." This idea showed for the first time how complex functional structures could have arisen by purely natural means, through chance rather than design. As a result, there are many who believe that Darwin's theory has destroyed once and for all the Argument from Design for the existence of God. That is why the philosopher Daniel Dennett calls natural selection "Darwin's Dangerous Idea," and made that the title of one of his books.[1] And it is why the biologist Richard Dawkins has said that Darwin, by giving us "design without design," made it possible for the first time to be "an intellectually fulfilled atheist."[2] Darwin, he says, provided an answer to Paley's famous watch argument.

If we think of the watch in Paley's argument as standing for the organic structure of a living thing, then, says Dawkins, we now know who made that watch: it was not God or any other intelligent designer; rather, it was the very universe itself acting blindly according to chance and the laws of physics. The universe is "the Blind Watchmaker"—the title of Dawkins's best-known book.[3]

If we look at a fish, for example, we observe that its fins, its streamlined shape, and many other features of its anatomy help it to swim better. It seems well "designed" for swimming. But that, according to Darwin, is not due to foresight on the part of any designer. No one looked ahead and, anticipating that a fish would swim, decided to incorporate these features into its design. Rather, it is thought that at some early time there existed aquatic creatures that were less sophisticated in structure and that were not such efficient swimmers. By a process of trial and error, various small changes in structure led the descendants of those creatures to be better and better adapted to this form of locomotion. Perhaps a creature with no fin had, due to random genetic mutations, offspring that had small fin-like protuberances. This novelty may have given those offspring a slight edge in the struggle for survival. Members of their species who had small fin-like protuberances consequently left more descendants on average than those who did not, and came to predominate. Over time, the same process of random mutation and natural selection led to creatures with larger and more functional fins.

Therefore, fins do indeed have a "purpose" in the sense of a natural function for which they are adapted, but they do not reveal a purpose in the sense of an intention that existed ahead of time in somebody's mind.

Thus, as Dawkins admits, the bodies of living organisms do have characteristics—complexity and functionality—that one normally associates with things that have been designed. But in light of Darwin's idea of natural selection we now realize, he says, that these structures arose in a completely different, haphazard way. He calls such objects, which appear to be designed, but really are not, "designoids."[4] Dawkins claims that living things, including human beings, are designoids. It was, he argues, due to the fact that people were familiar with designed objects, but did not understand the theoretical possibility of designoids, that they fell into the fallacy of the Design Argument for the existence of God. We shall return to these arguments in chapter 13.

While we are on the subject of biological structures, it should be mentioned that there is one huge question about life that natural selection does not seem able to answer, and that is how the first living thing originated. This is the so-called origin-of-life problem. It is generally believed by biologists that all living things on Earth are descended from a single original one-celled creature. How did that first creature evolve? It cannot have been by the ordinary Darwinian mechanism of natural selection, since for natural selection to operate there already has to be life—that is, self-reproducing organisms able to pass on their traits genetically.

The origin-of-life problem is made very hard by the fact that that first, "primitive" life-form was probably already enormously complicated. Partly in response to claims that vestiges of one-celled life had been discovered in Martian rocks, biologists have given some thought to the minimum requirements for a self-reproducing one-celled organism. It appears that it needs to have quite an elaborate structure, involving dozens of different proteins, a genetic code containing at least 250 genes, and many tens of thousands of bits of information.[5] For chemicals to combine in random ways in a "primordial soup" to produce a strand of DNA or RNA containing such a huge amount of genetic information would be as hard as for a monkey to accidentally type an epic poem.

Decades of attempts to solve this problem have borne little fruit.[6] Consequently, there are those who argue that none of the three ways of naturally explaining structure that we have discussed—pure chance, law, and natural selection—are capable of explaining the origin of life. They contend that this impossibility adds up to an irrefutable proof of divine intervention. However, this proof has a loophole. It is probably correct to say that natural selection and law are insufficient to explain the origin of life. But actually pure chance may be enough. The point is that the universe may well be infinitely large and have an infinite number of planets. This would be the case, for example, in the standard Big Bang cosmological model if the universe has an "open" geometry. (See chapter 7, pp. 48–52.) No matter how small the probability of life forming accidentally on a single planet, as long as that probability is finite, life is bound to form if there exist a truly infinite number of planets. Returning to the monkey-and-typewriter analogy: if there are an infinite number of monkeys typing, then some of them inevitably *will* be typing *Hamlet*. In fact, as remarkable as it sounds, it can be shown that if there are truly an infinite number of planets then somewhere there must exist a planet identical to ours in every particular, on which live species exactly like our own, and indeed on which lives a being who is genetically and in every other way biologically identical to you. Infinity is a very large number.

Are there an infinite number of planets? It is impossible for us to know by direct observation, because we cannot see what lies beyond our "horizon" of about 15 billion light-years. (See pp. 54–57.) But it is interesting that in order to explain the origin of life from inanimate matter in a way that does not invoke divine intervention it may be necessary to postulate an unobservable infinity of planets. Does this ring a bell? In chapter 7 we saw that in order to avoid a beginning of time, the materialist has to assume that before the Big Bang there was an infinite stretch of time about which (almost certainly) nothing can be known by direct observation. We shall see in later chapters other cases where the materialist, in order to avoid drawing unpalatable conclusions from scientific discoveries, has to postulate unobservable infinities of things. How ironic that, having renounced belief in God because God is not material or observable by sense or instrument, the atheist may be driven to postulate not one but an infinitude of unobservables in the material world itself!

In any event, we have identified three ways of explaining complicated structures naturally: (1) pure chance (combined, possibly, with the supposition of an infinite universe), (2) the laws of nature, and (3) natural selection. Does the possibility of such explanations fatally undermine the age-old Argument from Design? Has science, in this case at least, eroded the credibility of the religious worldview? It is to this question that I will now turn. The two versions of the Argument from Design that we distinguished between, the cosmic and the biological, raise somewhat different issues and should be discussed separately. I will begin with the Cosmic Design Argument and in particular with the question of whether the laws of nature can themselves be a substitute for God in explaining the order that is seen "in the heavens and on earth."

11 The Design Argument and the Laws of Nature

There is something paradoxical in the long-running debate over design. On the one hand, we have the older view that the lawfulness of the universe implies the existence of a lawgiver. This is the perspective of the Bible and of ancient religious writers, as we saw from the passages from Jeremiah, Psalms, and Minucius Felix quoted in chapter 9. It was also the view of many pagan writers of antiquity. On the other hand, we have many modern atheists who claim that the laws of nature prove the very opposite: that there is no need of God. One thinks of the famous statement of Pierre-Simon Laplace, who when asked by Napoleon why God was nowhere mentioned in his great treatise on celestial mechanics, replied, "I have no need of that hypothesis." (When told of this by Napoleon, another great physicist, Joseph-Louis Lagrange, is reported to have said, "Ah! But it is such a beautiful hypothesis. It explains many things."[1]) Are the laws of nature evidence for God or an alternative to "that hypothesis"? Do they prove that God is needed or that he is not needed?

At work here are two very different ways of thinking about the order found in nature. The old Argument from Design is based on the commonsense idea that if something is arranged then somebody arranged it. The reasonableness of this idea can be seen from an everyday example. If one were to enter a hall and find hundreds of folding chairs neatly set up in evenly spaced ranks and files, one would feel quite justified in inferring that someone had arranged the chairs that way.

One can imagine, however, that a person might object to this obvious inference, and suggest instead that the chairs are merely obeying some Law of Chairs operating in that hall. Most people, I suspect, would regard that as an absurd suggestion. Of course, it is true that the chairs in the hall are obeying a "law," in the sense that the positions of the chairs follow a precise mathematical rule or formula. But that is just another way of saying that the placement of chairs exhibits a pattern. The "law" in question does not *explain* the pattern, it merely *states* the pattern. It does not tell us why the pattern is there.

It is important to distinguish different kinds of "laws." Some laws cannot help but be true, such as the law of non-contradiction in logic, which says that a proposition cannot be both true and false, or the Fundamental Theorem of Arithmetic, which says that every whole number can be factorized in a unique way into primes. But the "Law of Chairs" is very different, because it does not express some necessary relationship. It is true that in the particular hall in which we found the chairs so precisely arranged there is a mathematical formula that can be used to predict where the next chair is in a row or column. That is, the Law of Chairs works. But it did not *have* to work; it could have been otherwise. The chairs might just as well have been positioned in some other way. And this is one reason why in real life we would conclude from the orderly arrangement of the chairs that someone *chose* that arrangement. In such a case we would argue that the law was due to a lawgiver, whereas we would not say that of the fundamental theorem of arithmetic or the law of non-contradiction in logic.

The critical point is that the laws of nature are not like the laws of logic or the laws of arithmetic, which have to be true. Rather, the laws of nature are simply patterns which we discover empirically in the world around us, but which could have been otherwise. In that sense they are just like the pattern found in the chairs in the hall, and therefore, the religious believer says, they must be the product of a mind.

It would seem, then, that the atheist is making an absurd argument when he says that the laws of nature have some kind of ultimate status that would permit one to dispense with a lawgiver or designer. However, there is really more to the atheist's point of view than would appear from this chair analogy. The atheist has a point to make which is worthy of serious consideration. To see what it is, it would help to consider a different analogy.

Let us imagine that we have a cardboard box with some marbles rolling around in the bottom of it. The marbles will tend to have a rather random distribution and roll around aimlessly. But, if we tilt the box slightly the marbles will all roll into one corner and we will see a pattern emerge—at least if we do a little jiggling of the box so that the marbles settle as much as they can. The pattern that will emerge is called the "hexagonal closest packing" pattern. It is the same pattern one sees in a honeycomb, or in the way oranges are sometimes stacked at a fruit stand (see figure 1).

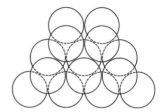

Figure 1. The most compact way to pack spheres of equal size is called a "hexagonal closest packing" arrangement. Marbles at the bottom of a tilted box will tend to arrange themselves like this to minimize their gravitational energy.

In this example the marbles really do seem to be obeying a Law of Marbles in some more meaningful sense than the chairs in the hall were obeying a Law of Chairs. Whereas the chairs might have been found positioned in many different ways, the marbles in the tilted box have no choice: there are physical and geometrical considerations that force them to be arranged in a certain way. Gravity is pulling the marbles downward, and consequently the marbles tend to get squeezed into the tightest arrangement that is geometrically possible.

In this example, physical forces (gravity) and mathematical necessity (the closest way to pack spheres) combine to produce a simple pattern. This would appear, then, to justify the claim that order can spring forth "spontaneously" and "necessarily" from disorder by unconscious laws rather than by intelligent design. If this is so, then laws of nature can truly explain order rather than merely describe it.

This is how many atheists look upon natural law in relation to the universe. And, to be sure, there are many things in nature that seem to bear out this view. Indeed, the whole history of the cosmos seems to tell the same tale. Gravitation made the chaotically swirling gas and dust that filled the universe after the Big Bang condense into stars and planets. The same forces made those stars and planets settle into orderly arrangements like our solar system. On the surfaces of planets, chemicals clumped together under the influence of electromagnetic forces into smaller and then larger molecules, until at last molecules that could replicate themselves appeared, and biology began. This is the grand picture: order emerging spontaneously from chaos. And presiding over the whole drama, the atheist tells us, is not some intelligence, but blind physical forces and mathematical necessity.

While this history of the cosmos is undoubtedly correct, the lessons the atheist draws from it are based on a superficial view of science. It is a view that really leaves out a major part of what science has taught us about the world, perhaps the most important part.

The overlooked point is this: when examined carefully, scientific accounts of natural processes are never really about order emerging from mere chaos, or form emerging from mere formlessness. On the contrary, they are always about the unfolding of an order that was already implicit in the nature of things, although often in a secret or hidden way. When we see situations that appear haphazard, or things that appear amorphous, automatically or spontaneously "arranging themselves" into orderly patterns, what we find in every case is that what appeared to be amorphous or haphazard actually had a great deal of order already built into it. I shall illustrate this first in the simple example of the marbles in the box, and then later in more "natural" cases like the growth of crystals and the formation of the solar system. What we shall learn from these examples is the following important principle: *Order has to be built in for order to come out.*

In fact, we shall learn something more: in every case where science explains order, it does so, in the final analysis, by appealing to a greater, more impressive, and more comprehensive underlying orderliness. And that is why, ultimately, scientific explanations do not allow us to escape from the Design Argument: for when the scientist has done his job there is not less order to explain but more. The universe looks far more orderly to us now than it did to the ancients who appealed to that order as proof of God's existence.

In Science, Order Comes from Order

Before we look at examples taken from nature, let us revisit the example of marbles in a box. We saw that if we tip the box slightly the marbles tend to form an orderly arrangement. But why does that happen? If I were do a similar thing with my living room—if I were to hire a huge crane to come and tilt it so that everything slid into a corner—I would not end up with an orderly pattern. I would find, instead, that the lamps, the furniture, the toys, and so on, would pile up into a jumbled heap. Why, then, don't the marbles form a jumbled heap? Part of the reason is that, unlike the objects in my living room, the marbles all have exactly the same size and shape. But this is not the whole story. After all, if I were to put a lot of identical spoons, say, in the bottom of a box and tilt it, the spoons would still form a jumbled heap. A crucial fact is that the marbles not only all have the same shape, but that that shape is a particularly simple and symmetrical one: the sphere. In fact, the sphere is the most symmetrical three-dimensional shape possible, because it looks exactly the same from any angle. So when the marbles fall into the corner it does not matter very much how they fall. Spoons or furniture pointing every which way will look like a jumble. But spheres cannot point every which way, because no matter which way a sphere is turned it looks just the same.

We see, then, that even before the box was tilted and the marbles lined up, there were two principles of order already present and at work: (1) every marble had the same size and shape as every other marble, and (2) each marble had the perfectly symmetrical shape of a sphere. And these principles of order are, in a basic way, very much like the principles of order that the chairs in the hall obeyed in our first example. When we found two neighboring chairs in the hall a certain distance apart, we knew that any other pair of neighboring chairs would be the *same* distance apart. And we also knew that if one pair of neighboring chairs were lined up in a certain direction, then any pair of neighboring chairs would be lined up in the *same* direction. In the case of the marbles, we know that if one marble has a certain size and shape, any other marble in the box will have the *same* size and shape. And we also know for each individual marble that if it looks a certain way from one direction it will look just the *same* from any other direction. All of these statements have to do with something being the *same* as something else, and as we shall see later they all have to do with various kinds of symmetry. In fact, the word *symmetry* is Greek for "same measure." Both the arrangement of chairs and the properties of the marbles exhibit what I called "symmetric structure" in chapter 9.

So why are the marbles in the box symmetric? Why are they all identical and spherical? Is it because of some ultimate Law of Marbles? Not at all. We know exactly why the marbles are symmetric: they were designed that way. Someone chose that they should be and had them manufactured accordingly. So we are back to where we started: the commonsense idea that design comes from a designer. At first it looked like the marbles-in-the-box example pointed toward a different conclusion. It seemed that the orderly arrangement of marbles could be traced ultimately to physical laws (gravity) and mathematical necessities (the closest way to pack spheres). But such explanations, while perfectly true, were *not* ultimate explanations. They presupposed an orderliness even greater than that which they explained: they presupposed the orderly or symmetric properties of the marbles themselves. And that orderliness and symmetry, we know, came not from mathematical necessity or physical laws but from design.

Why were we misled at first in thinking about the marbles in the box? Why did we think we could do without design? We were misled by a very natural mistake, a mistake that all atheists make in thinking about nature: *we took something for granted*. We took for granted the characteristics of the marbles and focused solely on their arrangement in the box as if that were the most fundamental fact and the only fact that called for explanation. But further thought showed us that the thing we took for granted requires explanation every bit as much as the effect we set out to explain.

IN SCIENCE, ORDER COMES FROM GREATER ORDER

We have seen in this simple example of the marbles how scientific explanations of order work in practice. They take order that is observed at a more superficial level (the hexagonal pattern of marbles in the bottom of the box) and show it to be the consequence of an order that is *presupposed* at a more fundamental level (the sameness and spherical shape of the marbles themselves). That is, when order comes out it is only because order was put in; so we still have order to explain. And—in the marble example, at any rate—the order that was put in was put in by a designer: the designer of the marbles.

Maybe there is a way out for the atheist, however. Perhaps, a scientific explanation could explain a lot of order by presupposing only a little bit of order. And maybe that little bit of order could be explained by a more fundamental theory that presupposed even less order. And so it might continue until one got to some ultimate and deepest theory that had to presuppose almost no order at all. In other words, at the end of the day, perhaps science will show that starting with some very trivial or negligible amount of order, the immense amount of order we see in nature might be built up by successive layers of explanation. And perhaps the order that exists at that deepest and most fundamental level will turn out to be so trivial that we do not have to attribute it to a designer, but can dismiss it as an accident.

This is certainly a hypothetical possibility, but it does not seem to be borne out by any actual examples. In fact, both in our simple marble example and in examples taken from nature, one finds quite the reverse happening: the order that is presupposed by a scientific explanation is *greater* than the order it accounts for. As one goes deeper and deeper into the workings of the physical world, to more and more fundamental levels of the laws of nature, one encounters not ever less structure and symmetry but ever more. The deeper one goes the more orderly nature looks, the more subtle and intricate its designs.

I have just asserted that at the deeper levels of nature one encounters more, rather than less, order. In order to justify such a statement, there must be some way of quantifying the amount of orderliness that a pattern or structure has. In the case of the precise kind of orderliness that is called by mathematicians "symmetry," it is indeed possible to do this, as we will now see.

How Much Symmetry?

Consider a small part of the hexagonal pattern of marbles that results from tipping the box. In particular, consider a little part consisting of seven marbles, as shown in figure 2.

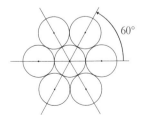

Figure 2. Spheres arranged in a hexagonal array ("snowflake pattern"). This array has a six-fold symmetry because rotating it by 60 degrees or any multiple of 60 degrees leaves it "invariant," i.e., looking the same.

Let us call this "the snowflake" for short. The degree of orderliness of the snow-flake can be measured by its degree of *symmetry*. In what way is the snowflake "symmetric"? For one thing, we can see that the pattern would look just the same if the snowflake were rotated 60 degrees clockwise. To say that it would look "just the same" is to say it has a symmetry. To a mathematician or physi-cist "symmetry" is defined as follows: A symmetry of an object is some operation that can be done to that object that leaves it looking the same as before, i.e., that leaves it "invariant." In the case of the snowflake, rotating it by 60 degrees leaves it invariant, and so the mathematician says that the act of rotating the snowflake by 60 degrees is a symmetry of the snowflake.

In fact, by this definition one sees that the snowflake possesses several sym-metries. There are actually six different rotations that can be done to the snow-flake that leave it looking the same: namely, rotations by 60 degrees, 120 degrees, 180 degrees, 240 degrees, 300 degrees, or 360 degrees. This means that the snowflake pattern has at least six symmetries. Actually, though, it has even more, because the act of flipping the snowflake upside down also leaves it invariant. Since one can combine the acts of rotating and flipping, it turns out that alto-gether the snowflake has a grand total of twelve symmetries. A mathematician would say that the snowflake has a "group of symmetries" consisting of twelve "elements." This group of twelve symmetries has a name; mathematicians call it D_6, one of the "dihedral symmetry groups." The snowflake example illustrates how one can count or measure the amount of symmetry that an object or pat-tern has.

Now, I said that the underlying reason that the marbles arranged themselves in this symmetrical pattern was that they all have the same size and shape and are all spherical. Let us consider this more closely. If one took the snowflake and turned it by 60 degrees, what exactly would happen? Except for the marble in the middle, each marble would move to a place in the pattern where another marble had been before. Therefore, unless the individual marbles were indis-tinguishable from each other the snowflake would not look the same after it was turned: the process of turning the snowflake would replace each marble, except the one in the middle, by a marble of different shape (see figure 3). Having marbles of different shapes or sizes would consequently spoil the symmetry.

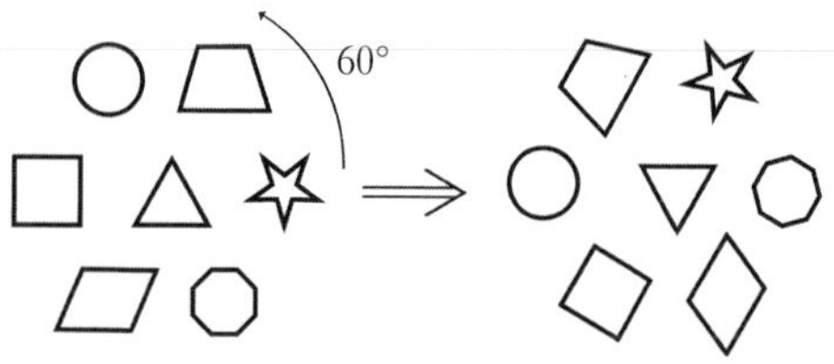

Figure 3.   Unlike the snowflake pattern in figure 2, a hexagonal array of six dissimilar objects does not have a six-fold symmetry, because rotating the array by 60 degrees makes it look different.

Another thing happens when the snowflake is turned by 60 degrees: in the process, each individual marble (including the one in the middle) is turned by 60 degrees about its own center. Thus, unless each marble had a shape (such as a sphere) that made it look the same after rotating it about its center, the snowflake would not end up looking the same, and the symmetry would be spoiled. For example, in figure 4, all the marbles in the snowflake are cubes, so turning the snowflake by 60 degrees does not leave it invariant and is therefore not a symmetry.

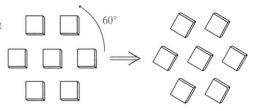

Figure 4. Even a hexagonal array of six similar objects will not have a six-fold symmetry unless each of the objects has a six-fold symmetry. For instance, if the objects are cubes, as shown here, rotating by 60 degrees makes the array look different.

So the symmetry of the snowflake ultimately arises from the properties of the marbles, namely that they are all the same in shape and size, and that their shape is spherical. And these properties are really themselves *also* symmetries. This is most obvious when it comes to the spherical shape of the marbles. A sphere is symmetrical in the same way that the snowflake pattern is symmetrical: it looks the same when one rotates it. However, a sphere is much *more* symmetrical than a snowflake. The snowflake only looks the same if it is rotated by 60 degrees, 120 degrees, 180 degrees, 240 degrees, 300 degrees, or 360 degrees (combined possibly with a flip), whereas the sphere looks the same if it is rotated by any angle whatsoever. Thus, while the snowflake had twelve symmetries, the sphere has an infinite number of symmetries. In fact, one could say that the sphere has a doubly infinite number of symmetries, because a sphere looks just the same if rotated by *any angle* about *any axis*. The doubly infinite "group" of symmetries of a sphere is called by mathematicians SO(3), the "rotation group in three dimensions."

Not only are there infinitely more symmetries of a sphere than of a snowflake, but, as we have seen, the symmetries of the sphere *include* all the symmetries of the snowflake. Therefore, it is true to say that the snowflake arrangement of marbles in the box is manifesting or exhibiting just a small part of the symmetries that were already built into the structure of the marbles themselves.

So far I have shown that the spherical shape of the marbles is an example of symmetry. But it is also the case, though not as obvious to a non-mathematician, that the indistinguishability of the marbles — the fact that they all have the same shape and size — is an example of symmetry. Suppose one has seven marbles of identical shape and size, not necessarily stuck together as they are in the snowflake, but even lying separately. One could take any two of those marbles and interchange them and it would make no difference; they would look just the same. So the act of interchanging two marbles is a symmetry. In fact, one could also interchange three or more of the marbles at a time. For instance, one could simultaneously replace marble A by marble B, marble B by marble C, and marble C by marble A. Such a "permutation" (as it is called by mathematicians) is also a symmetry. How many interchange and permutation symmetries do seven identical marbles have? It turns out that the total number is equal to $7 \times 6 \times 5 \times 4 \times 3 \times 2 \times 1 = 5{,}040$. Mathematicians call this group of 5,040 symmetries S_7, the "seventh permutation group." To say that the seven marbles have these 5,040 permutation symmetries is equivalent to saying that they are all indistinguishable or interchangeable.

Thus, underlying the twelve symmetries of the snowflake pattern of marbles that we saw emerge at the bottom of the box was a vastly greater number of symmetries intrinsic to the marbles themselves: an infinite number of rotational symmetries and 5,040 permutation symmetries. I have chosen to talk about seven marbles for concreteness, but the same points would apply no matter how many marbles were in the box.

Naively, when the marbles arranged themselves into the hexagonal array, it looked like symmetry was springing out of thin air. We were getting symmetry from chaos. But in reality what was happening was that a very small part of the symmetry that was already engineered into the marbles was manifesting itself in a certain way in their arrangement. Symmetry was not springing out of thin air, it was coming from greater symmetry. When the marbles chose to take on a particular hexagonal pattern, there was in a certain sense a *reduction* in the amount of symmetry. This is a phenomenon that in physics is called "spontaneous symmetry breaking." Some, but not all, symmetries of the marbles are reflected in the pattern they form in space. The other symmetries are "broken" by the pattern. Symmetry is not being made, it is being broken. The idea of spontaneous symmetry breaking is one of the most important in modern physics. It is one reason for the fact that the deeper structure of nature has more symmetry than the more superficial layers.

Let us reiterate the main points. In the example of the chairs in the hall, it was obvious that the chairs could have been arranged differently, and so one inferred that the pattern was the result of choice, and that an intelligence made that choice. The pattern was there "by design." In the example of the marbles in a box, it seemed at first that the marbles *had* to line up in a very particular way because of natural forces and mathematical necessity. Thus one might have concluded that no designer was involved. But, when this marble example was considered more carefully, it became clear that the pattern of marbles emerged only because of some very special prior facts about the marbles. And, when thought of even more carefully, these prior facts about the marbles were seen to be in themselves symmetric patterns that did not have to be as they were but were in fact the product of design. And, finally, it was found that the symmetry that was engineered into the marbles by their designer was of a *higher* degree than the symmetry of the pattern one at first thought arose spontaneously.

An Example Taken from Nature: The Growth of Crystals

As anyone knows who has ever seen frost on a pane of glass, very beautiful patterns can form automatically under the right conditions. No one imagines that water molecules are arranged in these wonderful feathery shapes by a miracle, or supernatural intervention. It is a completely natural process. The same is true of the lovely crystals that form deep within the earth. These assemble themselves spontaneously, with each molecule or atom attaching itself in the right place in the crystal without anyone telling it where to go.

The way this happens is quite similar to what happens in our marble example. The marbles in the bottom of the box line up in a hexagonal pattern because, under the influence of gravity, they are trying to find the lowest possible positions. In other words, the marbles are trying to lower their total "gravitational potential energy." In a similar way, crystals form because the atoms or molecules are trying to lower as much as possible something physicists call "free energy." It turns out that under certain conditions of temperature and pressure this free energy is lowest in certain materials if the atoms or molecules are arranged in crystals.

In a crystal, the pattern in which the atoms or molecules are arranged is called the "lattice." There are many different kinds of crystal lattices found in nature. For example, the lattices of diamond, calcite, galena, and mica are all different. And just as one can use the mathematical language of symmetry to describe the orderly arrangement of the marbles, so every kind of crystal lattice has a specific "group of symmetries." A specialized branch of mathematics is devoted to studying these "crystallographic symmetry groups." Take, for example, crystals of diamond. A perfect diamond lattice has a group of forty-eight symmetries called by mathematicians O_h, or the "hexoctahedral group."

In the marble example, we saw that the orderly arrangement of the marbles was a consequence, ultimately, of (1) the exact sameness of all marbles, and (2) the spherical shape of each marble. In a similar way, the possibility of diamonds forming presupposes (1) the sameness of all carbon atoms, and (2) spherical symmetry. [A technical point should be made for the sake of experts. It would be an oversimplification to say that the carbon atoms themselves were spheres.

But there is a precise sense in which the laws of electromagnetism, which govern the behavior of the carbon atoms, and the physical space in which the carbon atoms are moving are spherically or rotationally symmetric.]

Just as in the marble example, the order or symmetry that is manifest in the diamond crystals is far less than the underlying order and symmetry that gives rise to it. For underlying the forty-eight hexoctahedral symmetries of a perfect diamond crystal lattice are a huge number of "permutation" symmetries of the carbon atoms (due to the fact that they are all exactly the same and thus interchangeable) and an infinite number of "rotational" symmetries of space and of the laws of electromagnetism.

Crystals are one of the more dramatic and impressive displays of orderliness that we see in the world around us. But when science attempts to understand this phenomenon it uncovers a much more impressive orderliness at the atomic level. Nor must we stop at the atomic level; we can trace this orderliness to its roots at even deeper levels of physics.

We can do this by asking why all carbon atoms are exactly the same. The marbles in our example are all the same because they were made in a factory to the same specifications: they are all the same *by design*. In the case of carbon atoms, however, we have a physics explanation. All carbon atoms are the same because they are all made up of identical electrons, identical protons, and identical neutrons, and because these particles act and are acted upon by forces that operate in exactly the same way everywhere in the universe.

That, of course, raises the question of why all electrons (or protons, or neutrons) are identical, and why the forces of nature act in the same way everywhere. The answers to these questions lie at a level deeper than atomic physics: the level of Quantum Field Theory. Quantum Field Theory is the mathematical framework used to understand fundamental particles, like electrons. It tells us that these particles come from "fields." A field is something that fills all of space and can vibrate in certain characteristic ways; and those field vibrations can propagate along like waves in a pond. In quantum theory waves and particles are two different ways of looking at the same thing, so electrons can be thought of as particles or ripples in the "electron field" that fills all of space. This helps explain why all electrons are exactly alike. A good analogy is this: whether I stick my finger into a pond over here or over there, the same kind of ripples will spread out in either case, because the water of the pond is the same everywhere. Similarly, all electron waves (which is to say, all electrons) are the same because the electron field that fills space is the same everywhere.

Now it might seem that I have disproved the very point that I have been trying to establish. I said that the symmetries of diamond crystals were a consequence of even greater symmetries at the atomic level, among which were the fact that all carbon atoms are exactly identical. But now I appear to have explained away this symmetry of the carbon atoms by looking at the subatomic level and appealing to Quantum Field Theory. Whereas the marbles had to be engineered to be the same, all electrons *have* to be identical, because Quantum Field Theory says so. But this, again, is a superficial view. One must ask *why* the electron field is the same everywhere in space. That it is so is a very powerful fact which has to be assumed in constructing our theories of physics in order to agree with what we see in the real world. It is not an absolute logical or mathematical necessity by any means. One could imagine that fields had different properties in different places. True, that would strike most particle physicists as an ugly thing to imagine, but who said that the world had to be beautiful?

The fact that the electron field has uniform properties throughout all of space is itself a statement that the electron field possesses a very large degree of symmetry, in fact a much greater degree of symmetry than is enjoyed by any specific set of electron particles or carbon atoms.

Thus we see the same result repeated again and again as we trace phenomena down through layer after layer to the deeper levels of the world's structure. The symmetries and patterns found at one level are manifestations of greater symmetries and more comprehensive patterns lying concealed at the more fundamental levels. In what sense more comprehensive? In this sense: the symmetry we see in a diamond has to do only with diamonds, whereas the symmetries of the carbon atoms apply to all carbon atoms, whether in a diamond, a lump of coal, a piece of charcoal, or a pencil point. Similarly, the exact indistinguishability of carbon atoms is just a fact about one particular kind of atom. But the underlying fact about electrons applies to all electrons, whether in carbon atoms, or oxygen atoms, or anywhere else.

Someday—perhaps quite soon—we may have a yet deeper theory that explains why the electron field is the same everywhere: maybe superstring theory, or "M-theory," or some as yet undreamt-of theory. But we can be sure that whatever new and deeper theory comes along, it will reveal to us more profound principles of order and greater and more inclusive patterns.

What science has shown us is that most of the beauty and order in nature is hidden from our eyes. When ancient writers like Minucius Felix praised the "providence, order, and law in the heavens and on earth," they spoke about something of which they could have only fragmentary glimpses. If we look at the world around us, we see only hints of order here and there peeking out from amidst a great deal of apparent irregularity and haphazardness. But science has given us new eyes that allow us to see down to the deeper roots of the world's structure, and there *all* we see is order and symmetry of pristine mathematical purity.

THE ORDER IN THE HEAVENS

I have been trying to show how the order which we see in nature at one level has its roots in a more mathematically perfect order that exists at a deeper level. This point is so important that I think it is worth illustrating with another example. A particularly noteworthy example, from the viewpoint of both religious and scientific history, is provided by the highly orderly movements of the heavenly bodies. As we saw, this "order in the heavens" was one of the pieces of evidence most frequently cited in antiquity that there is design in the structure of the universe, and therefore a designer. And it was the attempts to understand these astronomical patterns—by such men as Copernicus, Galileo, Kepler, and Newton (all religious men)—that ultimately gave rise to modern science.

When ancient astronomers looked at the heavens, they saw the stars, the Sun, and the Moon traveling around Earth with what appeared to them to be perfectly uniform circular motions. This, naturally, impressed them. The ancient Greeks were especially impressed, for they knew quite a bit of mathematics, and knew that the circle is a very special geometrical shape. One thing that is special about it, as I have already explained, is that it is highly symmetrical. In fact, a circle is more symmetrical than any other geometrical figure that can be drawn on a flat surface. Circles have other beautiful mathematical properties as well, many of which were known to the Greeks, who therefore regarded it as especially fitting that the heavenly bodies moved in circular paths.

The planets, on the other hand, seemed to move in a more irregular way than the Sun, the Moon, and the stars. (Indeed, the word *planet* means "wanderer.") However, even the motions of the planets were understood as being made up of circles in Ptolemy's theory of "epicycles." All of this celestial harmony is in great contrast to what is observed to happen on Earth, where objects do not move in nice neat circles, but behave in exceedingly complicated and unpredictable ways. Moreover, the heavenly bodies never seem to undergo change or corruption, while on Earth everything eventually disintegrates or decays. These contrasts gave rise to the belief that celestial bodies were essentially different from terrestrial ones, and were even made up of an entirely different kind of matter.

Of course, most of these ideas have long since gone by the wayside. We now realize that, except for the Moon, the heavenly bodies do not go around Earth in circles, this being merely an illusion created by the rotation of Earth on its axis. What actually happens is that Earth and other planets travel around the Sun in orbits that are—as discovered in the seventeenth century by Johannes Kepler—elliptical in shape.

However, while some of the particular ideas of ancient astronomy have been shown to be oversimplified or simply wrong, the ancients were not at all wrong in seeing a great deal of orderliness in the heavens. In fact, this orderliness is greater than, and runs far deeper than, they ever imagined. Consider, for instance, Kepler's discovery that the planets' orbits are ellipses. This probably would have delighted the ancient Greek mathematicians, since ellipses were well known and much studied by them, and they were therefore aware that ellipses are mathematically beautiful shapes in their own right. An ellipse, like a circle, is an example of a "conic section." A conic section is called that because it is the shape of the cross section of a cone when it is sliced through at some angle. There are five kinds of conic sections: circles, ellipses, parabolas, hyperbolas, and straight lines. (Interestingly, these are exactly the five possible orbits that an object can trace out if it moves under the influence of the gravity of another body, according to Newtonian physics.) Many of the mathematicians of ancient Greece, including Menaechmus, Aristaeus, Euclid, and Apollonius of Perga, wrote lengthy treatises about the properties of conic sections.

Kepler discovered, in addition to the elliptical shape of the planets' orbits, two other "laws of planetary motion." One of these has to do with how a planet speeds up as it approaches the Sun and slows down as it moves away. Kepler found that a planet moves in just such a way that a straight line drawn between it and the Sun "sweeps out equal areas in equal times." The third of Kepler's laws was a precise mathematical relationship between a planet's distance from the Sun and the time it takes to complete one orbit. None of these three beautiful patterns discovered by Kepler in the motions of the planets were suspected to exist by ancient astronomers. Moreover, the Keplerian motions were far simpler and more mathematically elegant than the dizzying system of "cycles" and "epicycles" that had to be assumed in the older astronomy in order to accurately predict the motions of the planets. The order of the heavens as seen by the ancients fell far short of the reality found by Kepler.

In some of the details of their theories the ancient astronomers were not too far wrong. Although the planets do move in ellipses, these ellipses actually happen to be very close to being circles. Ellipses are ovals that come in a variety of shapes, from highly elongated to nearly circular. The amount of elongation is measured by a quantity called "eccentricity," which ranges from 0 to 1. A highly elongated ellipse has an eccentricity that is close to 1, while an almost circular ellipse has an eccentricity that is close to 0. It turns out that the eccentricity of the Earth's orbit is only 0.016. Neptune's orbit is even more circular, with eccentricity 0.008, while Jupiter's orbit has eccentricity 0.05, still quite close to circularity. Two other features of the ancient astronomical models turn out to be correct: the orbits of the planets all lie approximately in the same plane, and all of the planets go around in their orbits in the same direction.

There really is, then, as the ancients believed, a great deal of order and mathematical structure in the motions of the heavenly bodies. What we would like to know is, where does this structure come from? The answer, not surprisingly, is that it comes from the laws of physics. In fact, all of the features of planetary motion which I have mentioned can be explained quite straightforwardly from Newton's theory of gravity and Newton's laws of mechanics. But this does *not* mean that Newtonian science managed to explain away the "order in the

heavens." Quite the contrary. What Newton showed is that there is a more profound order, pervading *all* of nature, which reveals itself in a particularly transparent way in the celestial motions. Whereas Kepler's laws applied only to the
planets, Newton showed that they were a consequence of laws which governed
the motions of all bodies, whether in the heavens or on Earth, whether great
planets or little apples falling from trees. The grandeur of the orderliness of
nature discovered by Newton moved Einstein to write the following verses:

> Seht die Sterne, die da lehren
> Wie man soll den Meister ehren.
> Jeder folgt nach Newtons Plan
> Ewig schweigend seiner Bahn.

Einstein's biographer, Banesh Hoffmann, translates this:

> Look to the Heavens, and learn from them
> How one should really honor the Master.
> The stars in their courses extol Newton's laws
> In silence eternal.[2]

Of course, Newton would have felt that another master, infinitely greater than
himself, should be honored for what Einstein called "Newton's Plan."

In our earlier discussion we found that the symmetries that are evident upon
the surface of nature reflect even greater symmetries that lie deeply buried in
the underlying laws of physics. We can see this also in the present case. Consider the first of Kepler's laws, that planetary orbits are ellipses. It turns out that
this is a consequence of the fact that in Newton's theory of gravity the force of
gravity obeys what is called an "inverse square law." This is a very simple mathematical relation, which also, as it happens, is obeyed by the electrical force.
What it says is that the strength of the gravitational force varies inversely with
the square of the distance between the two gravitating bodies. (For example, if
one is three times as far away from the Sun the force of the Sun's gravitational
pull is only one-ninth as great.) If the strength of gravity depended in some other
way upon distance, much more complicated orbits than ellipses would result.
In most cases, in fact, the orbits would not be closed curves at all.

This inverse square law is a very special kind of law that results from the fact
that the carrier of the gravitational force, the so-called "graviton" particle, is
exactly massless. This masslessness of the graviton, in turn, is due to a very powerful set of symmetries called "general coordinate invariance" and "local Lorentz
symmetry," about which I shall have a little more to say in the next chapter. It
is not important for the reader to understand what these symmetries are, just to

know that the elegant elliptical shapes found by Kepler are only the tip of a huge iceberg of symmetric structure hidden in nature's laws.

Let us turn now to Kepler's second law, namely that planets "sweep out equal areas in equal times" as they go around the Sun. This fact can be shown to be a direct consequence of a very general principle of physics called the law of conservation of angular momentum. "Angular momentum" is a property of physical systems that tells how much they are rotating or swirling around. The principle says that under certain conditions the total amount of angular momentum a system has cannot change. The angular momentum of an object going around some center is given by the object's mass times its distance from the center times its speed around the center. If the object gets closer to the center, it has to speed more quickly around to keep its angular momentum the same; if its gets farther away, it has to move more slowly around. In fact, it turns out that it has to do this in exactly the way described by Kepler's equal-areas law.

The law of conservation of angular momentum is also a key ingredient in the explanation of why all the planets' orbits lie almost in the same plane, and why they are approximately circular. It is believed that the Sun and planets formed from the condensation of a huge cloud of gas and dust that was swirling around billions of years ago. As this cloud condensed, its temperature went up and it radiated some of this heat into the colder surrounding space. The cloud thereby lost energy, but it was unable to lose whatever angular momentum it started with. It is a straightforward mathematical exercise to show that if one drains energy away from a cloud of particles while keeping its total angular momentum the same, the cloud will assume a more and more disc-like shape. That is, all the particles will tend more and more to orbit in the same plane. One can also show that the orbits of individual particles will tend more and more to approximate circles.

There is another fact about the celestial motions that is explained by the law of conservation of angular momentum. The fact that Earth rotates with uniform speed on its axis—which is the reason for the heavenly bodies appearing to have a uniform circular motion around Earth—is due to the conservation of Earth's angular momentum.

We have just traced several of the regular patterns exhibited by planetary motion to a single law of physics, the law of conservation of angular momentum. But where does that law come from? It turns out that conservation of angular momentum arises, ultimately, from—what else?—*symmetries*. There is a very general theorem, proven in 1918 by Emmy Noether, which relates such "conservation laws" to symmetries of the laws of nature. Some examples of things that are conserved besides angular momentum are energy, ordinary momentum, and electrical charge. (There are many other examples, but they are somewhat esoteric.) All of these conservation laws are consequences of underlying symmetries.

The symmetries that are responsible for angular momentum conservation are of a kind I have already mentioned, namely "rotational symmetries" (the kind of symmetry a sphere has). What these rotational symmetries say is that the laws of nature do not distinguish one direction in space from another.

In sum, we have traced several patterns found in planetary motion to the fact that angular momentum is conserved, and that in turn to the fact that the laws of physics have rotational symmetries. We have traced other patterns of planetary motion to the inverse square law obeyed by gravitation; that in turn to the masslessness of the graviton; and that to underlying symmetries called general coordinate invariance and local Lorentz symmetry. But where do all *those* symmetries—the rotational symmetries, the local Lorentz symmetry, and so on—come from? Actually, we do not yet know, because we do not yet know what the deepest laws of nature are. However, there is no doubt that these symmetries of the presently known laws of physics have their roots in some still-greater symmetry or more profound principle of order that the as-yet-unknown fundamental laws of nature obey.

 12 Symmetry and Beauty in the Laws of Nature

In chapter 9, I distinguished between two kinds of structure, which I called symmetric structure and organic structure. Organic structure is generally found only in biological organisms or artificial machines, whereas symmetric structure abounds in every part of nature.

The notion of symmetric structure as defined in chapter 9 was very broad. It included not only things that have symmetry in the strict sense, but those which exhibit any kind of order, regularity, or pattern that can be mathematically described. Nevertheless, it is useful to concentrate on symmetry in the strict sense, because it is so precise a concept and because it plays such an important role in modern physics.

We saw a number of examples of symmetry in the previous chapter, including the twelve-fold symmetry of a snowflake pattern and the rotational symmetries of a sphere. However, there are many other kinds of symmetry than these, and a great variety of symmetries can be discovered in physical phenomena and in the laws of physics themselves.

To get a taste of the great richness of the idea of symmetry let us first look at just a few examples taken from a very restricted class of symmetries, namely symmetries of two- and three-dimensional geometrical shapes. I will start with these because such symmetries can be readily visualized. However, as we shall see later, the kinds of symmetries that are most important in fundamental physics are much more mathematically sophisticated and cannot be so easily visualized or even visualized by human beings at all.

Some Examples of Geometrical Symmetries

Example 1.   A straight line has a simple kind of symmetry called symmetry of translation. One of the meanings of the word *translate* is "transfer" or "change position." If one takes a straight line and translates it by sliding it along its own length by some distance, the line does not change. The individual points of the line, indeed, are shifted in space, but the line itself remains as it was before. Therefore, translating a line along its own length by some particular distance is a symmetry of the line. And since a line may be translated by any distance whatever, a straight line possesses an infinite number of such translational symmetries.

Example 2.   Rotating a circle about its center by a particular angle leaves the circle invariant and is therefore a symmetry. Since rotating by any angle is a symmetry of the circle, the circle has an infinite number of such rotational symmetries.

Example 3.   A helix has symmetries that combine translations and rotations. A helix is a spiral-staircase shape, which can be found in many places in nature, such as in the form of DNA molecules. If a helix is rotated around its axis by some angle and simultaneously moved along the axis by some corresponding distance, it will look exactly as before. This is why screws work. Since the threads of a screw are helical they can slide smoothly along themselves when the screw is turned as long as the screw moves along its length at the same time.

Figure 5.   A helix or spiral-staircase shape (shown here from the side) has symmetries that consist of rotating it about its axis by some angle and simultaneously sliding it a corresponding distance along its axis. Such a combined motion leaves the helix invariant.

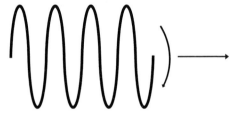

Example 4.   A slightly more subtle example of symmetry is found in the "logarithmic spiral," which is familiar as the shape of the nautilus shell. If such a spiral is rotated around its center, it appears to expand or contract. But if it is rotated by some angle and simultaneously magnified or reduced in scale by an appropriate factor, the logarithmic spiral remains unchanged. The symmetries of a logarithmic spiral thus involve both rotations and "scale transformations."

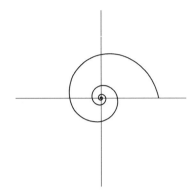

Figure 6. A logarithmic spiral (similar to the shape of a nautilus shell) has symmetries that consist of rotating it by some angle and simultaneously enlarging or reducing it by a corresponding scale factor.

Example 5. So far I have given examples of what are called "continuous symmetries." That is, each of the objects discussed has a continuous range of rotations, translations, or scalings which preserve its shape. Some objects, however, have only a countable number of "discrete symmetries." The snowflake pattern is an example; it has only twelve, as we saw before. A somewhat more intricate example of discrete symmetries is a soccer ball pattern. A soccer ball pattern is made up of twelve pentagons and twenty hexagons arranged in such a way that each pentagon touches exactly five hexagons and each hexagon touches exactly three pentagons and three other hexagons. This elegant pattern can actually be found in nature in a form of carbon molecule having sixty carbon atoms (C_{60}). The soccer ball shape can be shown to have exactly 120 rotational symmetries.

Figure 7. A soccer ball has a group of 120 symmetries, which are all the ways of rotating the ball that leave it invariant.

Example 6. Circles have symmetries other than the obvious rotational ones. They are also symmetric under "inversions." An inversion turns space inside out. It is done by taking a specific point in space as the "center of inversion" and then moving every other point in space directly away from (or toward) the center of inversion to a distance that is the reciprocal or "inverse" of its original distance, when measured in some units. For example, if the units are inches, all the points that were three inches away from the center of inversion become one-third of an inch away from it; the points one-seventh of an inch away become seven inches away; and so on. Most figures change their shapes when subjected to an inversion. But, remarkably, circles do not. Thus a circle has an "inversion symmetry." In fact, it has an infinite number of them, since it preserves its shape no matter what point in space is chosen as the center of inversion.

This example illustrates an important point. Symmetry is not always obvious. We can see that a circle has rotational symmetry just by looking at it. But for us to tell that it has inversion symmetry requires that we do some mathematical work. I imagine that this is due to the limitations of the human mind, and that to a more-than-human intellect the inversion symmetries of a circle might be as obvious as the rotational ones are to us. We shall see later in this chapter that physicists have uncovered many symmetries in nature that are very far from obvious and that require very sophisticated mathematical tools and concepts in order to understand them.

Example 7.   The connection between the ideas of beauty and proportion and mathematical symmetry can be nicely illustrated by a ratio called the "golden mean." This is the number $(1 + \sqrt{5})/2 = 1.618. \ldots$ Since ancient times artists and architects have seen in the golden mean the most aesthetically satisfying geometric ratio. It is claimed, for example, that the most pleasingly proportioned rectangle is one in which the ratio of height to width is the golden mean. Rectangles of other proportions, it is said, seem too narrow or too squat. Figure 8 shows a rectangle whose sides are in the ratio of the golden mean.

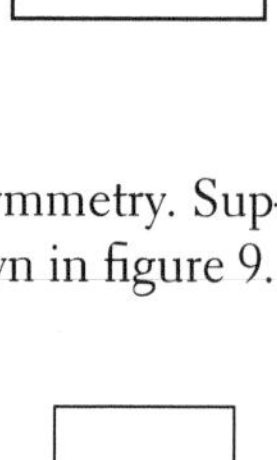

Figure 8.   A golden rectangle is one whose height and width are in the ratio called "the golden mean," considered to be the most pleasingly proportioned. The golden mean is the number $(1 + \sqrt{5})/2$, which is about 1.618.

It is not at all obvious, but this rectangle has a kind of scaling symmetry. Suppose that one slices off a square piece from the rectangle as shown in figure 9.

Figure 9.   A golden rectangle has a subtle symmetry. If it has a square piece cut off and is then rotated by 90 degrees and magnified, it looks exactly as it did to start with.

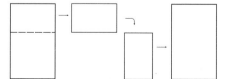

One is left with a smaller rectangle. This smaller rectangle is just a scaled-down version of the original rectangle—i.e., its sides are also in the ratio of the golden mean. In other words, if one cuts off a square, then scales what is left by an appropriate factor and rotates it by 90 degrees, one ends up with a rectangle that is identical to the original one. Under that procedure the "golden" rectangle is invariant. (If one starts with a rectangle wider than a golden one, then slicing off a square produces a rectangle thinner than a golden one, and vice versa.)

The golden mean is related to symmetry in other ways too. Consider the regular pentagon shown in figure 10. This has an obvious five-fold rotational symmetry. The ratio of one of the diagonals (shown as a dotted line) to one of the sides is the golden mean.

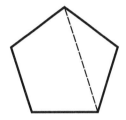

Figure 10.   The golden mean is related to the five-fold symmetry of a regular pentagon. The ratio of a diagonal (shown as a dotted line) to the length of a side is the golden mean.

In showing that there is a connection between beauty and symmetry, the example of the golden mean may not be the most convincing for everyone. However, there is no question that mathematical symmetry plays an important role in art. Geometrical symmetries, sometimes of great subtlety, can be found in all kinds of decorative patterns, such as friezes, tilings, arabesques, or French gardens. In architecture they are ubiquitous. One need only think of the eight-fold symmetry of the rose window, or the intricate patterns of the rib work, groins, and vaults in Gothic cathedrals. In music, mathematical symmetry is found in the very scales of notes. These scales do in fact have scaling symmetries. The mathematical nature of musical scales was first discovered by Pythagoras in the sixth century B.C. Notes which harmonize well with each other are those whose frequencies are in simple, precise whole-number ratios to each other. Symmetry is also found in the meter, rhythm, and melody of music. In poetry, symmetric structure exists in the metrical and rhyme schemes. In dance, it appears in the patterns of steps and sometimes in the patterns of the dancers themselves.

Symmetry contributes to the artistic unity of a work, to its balance, proportion, and wholeness. The connection between symmetry and unity is exceedingly important and applies also to symmetry in physics. If one part of a symmetrical pattern or structure is removed, typically its symmetry is spoiled. Cut off one arm of a starfish, and it will no longer have a five-fold but only a two-fold symmetry. Remove one pentagonal facet of a soccer ball, and instead of

120 symmetries only 5 will be left. Symmetry requires all the parts of a pattern to be present, and is therefore a unifying principle.

There is a famous and striking example of this in physics, where the incompleteness of a symmetric pattern actually led to the prediction and discovery of a hitherto unknown particle. It was known by the early 1960s that the properties of particles of the strongly interacting variety, known as "hadrons" (of which there are hundreds), exhibit certain precise mathematical regularities. Some of these regularities have to do with two important properties of hadrons called their "hypercharge" and "isospin." It is not important for the present discussion to know what hypercharge and isospin are, except that they are numbers that characterize the behavior of these particles in their interactions with each other. (For the experts: by isospin I mean more precisely the "third component of strong isospin.")

To take two well-known particles as examples, the proton has isospin $+\frac{1}{2}$ and hypercharge $+1$, while the neutron has isospin $-\frac{1}{2}$ and hypercharge $+1$. The proton and neutron belong to a set of eight closely related particles. The symbols used to denote these eight particles are $p, n, \Sigma^+, \Sigma^0, \Sigma^-, \Lambda, \Xi^0$, and Ξ^- (p stands for proton, n for neutron). That these eight particles are somehow related to each other can be seen if we plot the values of their isospin and hypercharge on a graph, as shown in figure 11. Here isospin is plotted along the horizontal axis and hypercharge along the vertical axis.

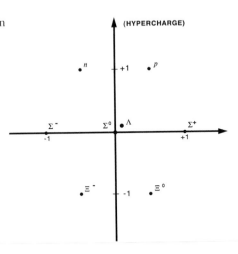

Figure 11. This graph displays certain properties of elementary particles. Each particle is represented by a dot whose horizontal position gives its "isospin" and whose vertical position gives its "hypercharge." The eight particles represented by these dots are closely related by "flavor SU(3)" symmetry and form a "multiplet." The dots labeled p and n represent the proton and neutron. The other dots, labeled by Greek letters, represent more exotic particles.

One sees immediately that these properties of the eight particles fall into a hexagonal pattern. Other properties of these eight particles also exhibit regularities. For instance, particles in the same horizontal row in figure 11 have masses that are nearly equal to each other: the masses of the proton and neutron are 938.3 MeV and 939.6 MeV, respectively. (*MeV* means "million electron Volts" and is a measure of energy or mass used in particle physics and nuclear physics.) The Σ^+, Σ^0, and Σ^- particles have masses of 1,189.4 MeV, 1,192.6 MeV, and 1,197.4 MeV, respectively. The Ξ^0 and Ξ^- particles have masses of 1,314.9 MeV and 1,321.3 MeV. A group of particles that have closely related properties like this are said to form a "multiplet." This particular multiplet is called an "octet."

The reason that hadronic particles come in such multiplets is the existence of a group of symmetries called "flavor SU(3) symmetry." The word *flavor* is a technical term, somewhat whimsically inspired by the idea that the different particles in a multiplet are like slightly different flavors of the same basic thing. As a result of these flavor SU(3) symmetries, all hadronic particles can be arranged in multiplets of different sizes. Some multiplets are "singlets" consisting of just one particle. Others are octets like the one we discussed. Others are larger. In 1962, one of these multiplets looked like this:

Figure 12. A multiplet of particles as it appeared in 1962. The mathematics of the SU(3) symmetry imply that there should be a tenth particle in this multiplet that completes the triangle. This particle, called the Ω^- (omega-minus), was predicted by Murray Gell-Mann in 1962 and was experimentally discovered shortly thereafter.

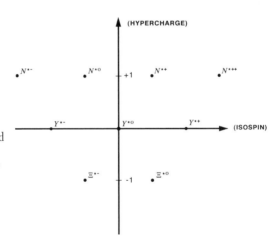

Without knowing anything about SU(3) symmetry, one could guess, just from the shape of the multiplet diagram, that there should be a tenth kind of particle down at the bottom to complete the triangle pattern. This is not just a matter of aesthetics—the SU(3) symmetries require it. It can be shown from the mathematics of the SU(3) symmetry group that the multiplets can only come in certain sizes. These include multiplets with 1, 8, 10, 27, and certain higher numbers of particles—but not 9. On the basis of SU(3) symmetry, Murray Gell-Mann predicted in 1962 that there must exist a particle with the right properties to fill out this "decuplet." Shortly thereafter the new particle, called the Ω^-, was indeed discovered.

What kind of symmetry is this "flavor SU(3)"? From the diagrams we have seen one might guess that it is the group of symmetries of a triangle or hexagon, i.e., that it consists of rotations by multiples of 60 or 120 degrees in space. But this is wrong. First of all, the rotations involved are not rotations in ordinary physical space. Note that the positions of the points in the diagrams do not represent the locations of particles in ordinary physical space. Rather, they graphically represent certain internal properties of the particles, specifically their "isospin" and "hypercharge." A flavor SU(3) "rotation," therefore, does not move a particle from one place to another in physical space, but moves it around in the diagram. In other words, it corresponds to a change in the *properties* of the particle. For example, it could turn a proton into a neutron, or a Σ^+ into a Ξ^0.

The symmetries we are most familiar with in everyday life are those of the geometrical shapes of objects, or symmetries involving time, like the symmetries connected with musical scales and rhythm. Such "space-time" symmetries do indeed play an enormously important role in physics. The flavor SU(3) symmetry, however, is an example of what physicists call an "internal symmetry," since it pertains to the inherent or internal properties of particles rather than to space and time. Even though the flavor SU(3) symmetries do not have to do with ordinary space, they can be thought of as having to do with an abstract space called "SU(3) space." One can think of the proton, neutron, and other particles as living simultaneously in ordinary space and in this SU(3) space. SU(3) space is rather a strange place. It has three "complex" dimensions. That is, positions in this SU(3) space are given by three coordinates that are not ordinary real numbers, but complex numbers. (A complex number is a number that can be written as $a + bi$, where a and b are ordinary real numbers and i is the square root of -1.)

The idea of an abstract complex-three-dimensional flavor space is not an easy thing to grasp. But that goes precisely to one of the important points that I wish to get across in this chapter, namely that the symmetries discovered in nature by modern physics are far more subtle and intricate than any that are encountered in ordinary experience or that were imagined in the past.

Another good example of symmetries of nature that are beyond our powers to visualize are those which are the basis of Einstein's Theory of Special Relativity. In popular expositions of relativity theory, and even in some textbook treatments, the idea of symmetry is not emphasized or even mentioned. However, in actuality, the entire content of relativity is contained in the statement that the laws of physics are Lorentz symmetric. The Lorentz symmetry group is a group of rotational symmetries. However, Lorentz rotations include not only rotations in three-dimensional space, but also rotations in four-dimensional space-time. There are two reasons we cannot directly visualize such rotations. The most obvious is that we are not biologically equipped to visualize more than three dimensions. The other reason is that space-time has geometrical properties that are fundamentally quite unlike those found in ordinary Euclidean space. Rotations in space "go around in circles," one might say. This would be true no matter how many dimensions of space one had. But rotations in space-time (i.e., Lorentz rotations involving both space and time) do not go around in circles; they go around in *hyperbolas*. (A hyperbola is shown in figure 13.) How can a rotation go around in a hyperbola? That, again, is just the point. One can explain it using abstract mathematics, but one cannot truly visualize it. At least, human beings, with our mental limitations, cannot.

Figure 13. This curve is a hyperbola. Rotations in space go around in circles, but rotations in space-time that involve both space and time go around in hyperbolas. Such rotations can be described by abstract mathematics, but cannot be visualized.

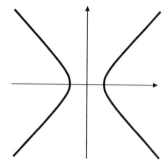

The connection between Lorentz symmetry and the usual discussions of relativity in terms of "frames of reference" is quite direct. Two observers (or laboratories) that are in uniform motion relative to each other are said to have different frames of reference. What this means mathematically is that the two observers or laboratories are oriented differently in space-time. The time direction of one reference frame is tilted at an angle in space-time with respect to the time direction of the other: that is, one frame is Lorentz rotated with respect to the other. Saying that both the laws of physics and the speed of light look the same in every reference frame—which is all that relativity says—is saying that they look the same from any angle in space-time. This is obviously a statement about symmetry.

I said earlier that symmetry in physics, as in art, is a unifying principle. This is readily seen in the case of the flavor SU(3) symmetries of particles. What seemed at first to be hundreds of different hadron particles turned out to be close relatives of each other. For instance, one could say that the eight particles in the octet containing the proton are in a real sense merely different facets of one symmetric entity. Since Einstein's theory of relativity is based on symmetry, one would expect that it too leads to some kind of unification, and indeed that is the case. In fact, it leads to several unifications. The most obvious is the unification of space and time. In Newtonian physics space and time were quite distinct, but in Einsteinian physics they together make up one symmetric four-dimensional space-time manifold. A second unification is of electric and magnetic fields, which the Lorentz symmetries reveal to be merely different facets of one entity called the electromagnetic field. A third unification is of mass and energy. These were distinct concepts in Newtonian physics, but were shown by Einstein to be intimately related. This is the meaning of his celebrated formula $E = mc^2$.

As remarkable as the flavor SU(3) and Lorentz symmetries are, they are only a small part of the story. Principles of symmetry are really at the heart of all of modern physics. For example, every one of the four basic forces of nature — gravity, electromagnetism, the strong force, and the weak force — is based in a profound way on principles of symmetry. The modern theory of gravity is Einstein's Theory of General relativity, proposed in 1916 and repeatedly confirmed since by experiment. General Relativity is based on symmetries called "general coordinate invariance" and "local Lorentz symmetry." Electromagnetism is based on something called "gauge symmetry." (This name was coined by Hermann Weyl in 1919. He was also the first to appreciate the fundamental importance of this symmetry.) In the 1960s and 1970s, physicists began to realize that the other two forces of nature — the strong and weak forces — are also based upon fundamental symmetries, and that these symmetries are mathematically similar to the "gauge symmetry" of electromagnetism.

When we say that all these forces are "based on" symmetries, we mean several things. Most profoundly, the very fact that there are such forces in nature is a consequence of these symmetries. If nature did not have these symmetries, it would also not have these forces. In addition, the characteristics of these forces are controlled by their symmetries. The structures of the mathematical laws governing these forces are to a large extent determined by their underlying symmetries. So much is this the case that modern fundamental physics is not so much driven by the search for new kinds of matter or new forces, but for the new and more powerful principles of symmetry that are suspected to lie beneath the surface of what is presently understood.

A good illustration of this came in the 1960s and 1970s. The discovery in that period that the three non-gravitational forces are all based on symmetries of the "gauge" type led to the realization that a mathematical unification of the theories of these forces could be achieved. The work of Sheldon Glashow, Abdus Salam, and Steven Weinberg, culminating in Weinberg's classic 1967 paper, showed how two of the forces — electromagnetism and the weak force — were unified. This "electroweak" theory has been dramatically confirmed in numerous experimental tests. In 1974, several physicists proposed theories in which the strong force was unified with the electroweak force in a "grand unified theory" of all non-gravitational forces. The idea of grand unification helps explain a number of important facts about nature, but so far has not been proven experimentally, even though there are several tantalizing hints in its favor.

The two most studied kinds of grand unified theory are based on symmetry groups called SU(5) and SO(10). SU(5) is related to rotations in a space with five complex dimensions. SO(10) is related to rotations in a space with ten real dimensions. The spaces in question, however, are not ordinary space, but abstract spaces somewhat like the "SU(3) space" we talked about before in connection with flavor SU(3) symmetry. Rotations in these SU(5) or SO(10) spaces would not move particles around in ordinary space, but would have the effect of changing their internal or intrinsic properties, for example turning an electron into a quark.

In the last twenty years the belief has grown among particle theorists that another kind of symmetry, called "supersymmetry," probably exists in nature. There are many theoretical reasons for this belief, but as yet there is no experimental proof. Supersymmetry is even more difficult to grasp intuitively than the Lorentz symmetries, the flavor SU(3), or the grand unified SU(5) and SO(10) symmetries. Those symmetries, at least, had to do with space-time or with abstract spaces that, although of higher dimensionality, could be described using ordinary numbers or complex numbers. Supersymmetry, however, has to do with spaces that must be described using bizarre quantities called Grassmann numbers. If x and y are ordinary numbers, or even complex numbers, then $x \times y = y \times x$. But if x and y are Grassmann numbers then $x \times y = -y \times x$. (A result of this is that any Grassmann number multiplied by itself gives zero. To see this, take x and y to be the same number in the above formula. Then one obtains $x \times x = -x \times x$. But this can only be so if $x \times x = 0$.) Needless to say, we do not encounter this kind of number in ordinary life, but they do naturally arise in describing certain kinds of particles, such as the electron.

Supersymmetry can be thought of as relating to rotations in a "superspace" that contains the ordinary four-dimensions of space-time as well as extra dimensions that require Grassmannian coordinates to describe. As wonderful as supersymmetry and grand unified theories are, they are not, in the view of most theorists, the end of the story. Many of the best minds in the field strongly suspect, and some are convinced, that the ultimate theory is one called "superstring theory," or "M-theory," which unifies all four forces of nature, including gravity. One reason for this suspicion or conviction is that superstring/M-theory is the only way known at present to formulate a theory of gravity that is consistent both with the theory of relativity and with quantum theory. The fundamental symmetries and mathematical principles on which M-theory is based are as yet only slightly understood. However, it is known that they are vastly richer and more subtle than those of any theory ever before studied.

13   "What Immortal Hand or Eye?"

THE ISSUE

In the foregoing chapters we have observed how, from ancient times, evidence for the existence of a cosmic designer has been seen in the structure of the universe, or, in the words of Minucius Felix, in the "order and law in the heavens and on earth." We saw that there are two kinds of structure. One, which we called "symmetric structure," is characterized by regularity, order, pattern, and symmetry. The other, which we called "organic structure," is characterized by a complex, functional interdependence of parts. We saw that the appearance of symmetric structure in natural phenomena can be explained using the laws of physics, as in the case of the growth of crystals or the formation of the solar system. Does this mean that science has succeeded in completely accounting for the fact that there is symmetric structure in the universe without needing to invoke a designer? It seems clear that the answer is no.

It is true that science explains order, but only by showing that it comes from some grander, more profound order, which it expresses in mathematical laws. The ultimate goal of physics is to uncover *the* laws of nature, for indeed most physicists share the belief that there do exist a set of truly fundamental laws, beyond which science cannot go. All scientific explanations would flow ultimately from those fundamental laws. We do not yet know what those laws are in their entirety, but we do know that the closer we have approached them the more profound and beautiful is the symmetric structure that has been revealed. In the words of an eminent theoretical particle physicist:

> Fundamental physicists are sustained by the faith that the Ultimate Design is suffused with symmetries.
>
> Contemporary physics would not have been possible without symmetries to guide us. . . . Learning from Einstein, physicists [look for] symmetries and see that a unified conception of the physical world may be possible. They

105

hear symmetries whispered in their ears. As physics moves further away from everyday experience and closer to the mind of the Ultimate Designer, our minds are trained away from their familiar moorings. . . .

The point to appreciate is that contemporary theories, such as grand unification or superstring, have such rich and intricate mathematical structures that physicists must marshal the full force of symmetry to construct them. They cannot be dreamed up out of the blue. Nor can they be constructed by laboriously fitting one experimental fact after another. These theories are dictated by Symmetry.[1]

It seems virtually certain that the ultimate mathematical laws of nature will turn out to have a beauty and depth of structure beyond anything seen so far. In fact, many theoretical physicists think they have these ultimate laws in sight. The "final theory," they suspect, is none other than the "superstring theory" mentioned in the passage just quoted, or, as it is now called, "M-theory."[2] The mathematical structure of this theory is so profound that after twenty years of intensive research by physicists and mathematicians, they feel that they have barely scratched its surface. Nevertheless, enough is known about the theory that experts marvel at it, using such words as *miraculous* to describe it. One of the greatest physicists of our time, in describing superstring theory to a layman, felt frustrated by his own inability to communicate the grandeur and magnificence of what his research had revealed to him: "I don't think I've succeeded in conveying to you," he said, "its wonder, incredible consistency, remarkable elegance, and beauty."[3]

The story of science is in many ways like the old fairy tale in which the hero is confined by a witch to the bottom of a dark well, but discovers there by luck and courage the secret entrance to her subterranean storehouse. Deep under the fields and forests he finds unimaginable treasures of rare beauty. Little did the early researchers in science, laboring over foul-smelling test tubes or experimenting with magnets and coils of wire, imagine what secret beauty lay at the bottom of the deep well of nature.

If the *ultimate* laws of nature are, as scientists can now begin to discern, of great subtlety and beauty, one must ask where this design comes from. Can science explain it? That is not possible. For if science always explains design by showing it to be part of or a consequence of a deeper and greater design, then it has no way to explain the *ultimate* design of nature. The *ultimate* laws of physics are the end of the road of scientific explanation. One cannot go any farther in that direction. Thus, if at the end of that road one is confronted with a magnificent example of what we called "symmetric structure" in the ultimate laws themselves, then science really has no alternative to offer to the Argument from Design.

Let us go back to the three possibilities we listed in chapter 10 for explaining structure naturally: chance, the laws of nature, and natural selection. Could natural selection explain the structure of the ultimate laws of physics? Some well-known physicists have speculated that perhaps in some way nature evolves its own laws. But this is clearly untenable. At least it is untenable if we mean a *natural* process of evolution. For any natural process is itself governed by natural laws. To say that the ultimate laws of nature evolved is, thus, self-contradictory. If a law evolves naturally, it must do so by a process governed by some other law, and therefore that other law would be more fundamental, and the law that evolved would not be the ultimate one.

Is it possible that the structure of the ultimate laws of nature can be explained by the ultimate laws of nature? Obviously, such an "explanation" would be circular.

We are left then with only one of the three natural explanations to explain the "order and law in the heavens and on earth": chance.

CAN CHANCE EXPLAIN IT?

We saw in chapter 10 that structures can arise by chance if there are sufficiently *many* chances for them to do so. If a monkey types one page of text at random, it is absurdly unlikely to produce even one grammatical sentence, let alone a poem in iambic pentameter. But give the monkey enough time—a time compared with which the entire age of the universe is the twinkling of an eye—and he might do it.

This means that if the structure of the laws of physics happened by chance, there must have been a great many chances. The only way one can imagine this being so is that ours is one of a great many universes, each with its own characteristics. Most of these universes would be highly irregular and disorderly, or would, if they had laws, have laws with little in the way of symmetry or beauty or simplicity. But a few could happen to be universes like our own. In the words of the philosopher David Hume, "Many worlds might have been botched and bungled, throughout an eternity, ere this system was struck out."[4]

Think of each universe as being one of the pages witlessly typed by the monkey in our analogy. Most of those universes will exhibit no pattern, or very little; they will be jumbles. But if there are enough such universes "typed," by pure chance a few might contain very elaborate and wonderful patterns.

This idea is very close to some versions of the Anthropic Principle, a topic that I shall discuss in a later chapter. However, in the form we are encountering it here the idea has a severe problem. If there are a vast number of distinct and different universes, it could be that by pure chance some of them would exhibit a high degree of mathematical structure, but that would *not* explain why we ourselves happen to live in one of these special, orderly universes.

Now, it could be argued, in answer to this objection, that only in a universe which had a lot of structure—as opposed to being utterly chaotic—could such elaborate structures as living beings arise; and this is doubtless a valid point. It probably does have to be the case that any complicated organism, such as ourselves, which was capable of complex mental processes, would necessarily find itself in one of those exceptional universes which had a great deal of order and structure in it. But that is not enough to explain the facts we see. We do not just live in a universe with a great deal of order. We live in a universe whose order is perfect, or nearly so, in the sense that exceptions to the mathematical rules hardly ever, or never, seem to occur.

If our universe is a page of text typed by a monkey, it is not just a page of text that more or less satisfies some grammatical rules. It is a flawless text, without grammatical or typographical errors. Why is that so? Certainly, life could have evolved just as it did even if there had been occasional lapses from the orderliness of nature. Such lapses might have a harmful effect on living beings, but they would not necessarily have to be any worse than various disasters that happen in accordance with natural regularities. An occasional small-scale violation of the law of conservation of angular momentum, say, would not necessarily be worse than a viral epidemic, or earthquake, or forest fire, as far as living things were concerned. Why, then, do we not see such violations of natural law? Why is the orderliness of nature *so* perfect?

To put it another way, if the universe is orderly and highly structured simply by the luck of the draw, then it is a miracle that miracles do not happen all the time. To use another analogy, of a thousand naturally formed gemstones, one would not expect to find any that were without numerous small flaws, parts that did not fit the crystal pattern perfectly. These might be mostly microscopic, but they would be detectable by a jeweler's instruments. Yet scientists have studied the physical universe with instruments of astonishing precision; and while they quite often find anomalous behavior that does not fit the laws of nature as they think them to be, it has always turned out that these anomalies could be accounted for by some *more* beautiful law. The universe does not appear more and more flawed the more closely one looks at it, as one might expect if its regularity were a matter of luck. Rather, its *fundamental* patterns appear more and more wonderfully perfect the more closely they are examined.

We have arrived at the point where we see that science can by no means explain away the rich design of nature and its laws. Science has only shown that design to be more magnificent than anyone had ever dreamt. Therefore, the Cosmic Design Argument for the existence of God still stands. Indeed, it is stronger than ever before. That, certainly, was the view of Hermann Weyl, one of the great mathematicians and mathematical physicists of the twentieth century:

Many people think that modern science is far removed from God. I find, on the contrary, that it is much more difficult today for the knowing person to approach God from history, from the spiritual side of the world, and from morals; for there we encounter the suffering and evil in the world, which it is difficult to bring into harmony with an all-merciful and all-mighty God. In this domain we have evidently not yet succeeded in raising the veil with which our human nature covers the essence of things. But in our knowledge of physical nature we have penetrated so far that we can obtain a vision of the flawless harmony which is in conformity with sublime reason.[5]

Is Natural Selection Enough?

Up to this point I have focused on what I called symmetric structure. But what about the other kind, organic structure, which is found in the biological realm, and which has also been regarded as evidence of a cosmic designer? As we saw, most biologists think that Darwin succeeded in explaining organic structure in a completely natural way. This has led some of them to claim that Darwin has exploded the Argument from Design. However, there is a point that cannot be emphasized too strongly: Even if these biologists are correct, and Darwin has explained the formation of biological structure, that would at most affect one version of the Design Argument for the existence of God, namely the Biological Design Argument. It would leave completely untouched the Cosmic Design Argument, which takes as its starting point the structure of the universe as a whole. I have argued that this structure, and in particular the structure of the laws of physics, cannot, in the final analysis, be explained by some kind of theory of natural selection.

Having said that, let me ask a scientific question: Is it indeed the case that natural selection can satisfactorily explain the complexity of structure that we see in the biological world? The great majority of scientists would probably unhesitatingly answer yes, and quite possibly they are right. Certainly I cannot prove that they are wrong. Therefore, for the sake of argument, I am going to make the assumption throughout this book that natural selection is a sufficient explanation, even though personally I am of the view that it is a completely open question scientifically.

Perhaps I should explain, before proceeding, why I think it is an open question. Basically, the reason is that there is just not enough evidence to settle the issue one way or the other at the present time. There is a great deal of evidence—it seems to me to be overwhelming evidence—that evolution happened. What I mean by that is that there is a great deal of evidence pointing to the fact that all life on Earth evolved from a common ancestor. There is also a great deal of evidence that natural selection plays a large role in evolution. What is lacking is sufficient evidence to prove that natural selection by itself is capable of doing the whole job of driving evolution. On top of that, there are some discoveries in recent decades that make that job look a lot harder than it once did.

It used to be thought that natural selection had virtually infinite stretches of time to work its wonders. After all, Earth is about 4.5 billion years old. However, it has been discovered that about 540 million years ago a huge number of complicated types of organisms quite suddenly appeared from much simpler types. So suddenly did this happen, in fact, that it is called the "Cambrian Explosion." It seems that this evolutionary leap took less than 5 million years to complete. That is about a hundred times faster than anyone dreamt evolutionary changes of that magnitude could happen by natural selection.[6]

It is not only the shortness of the time scales that gives one pause, but the incredible degree of complexity that must be accounted for. The simplest living thing, a bacterium, has a genetic blueprint that contains hundreds of thousands of "bits" of information. It is often said that the information required to put together a bacterium is about what is contained in the *Encyclopedia Britannica*. The molecular machinery inside a single cell of a plant or animal is so involved as almost to defy description. It has been argued by the molecular biologist Michael Behe that some cellular processes are "irreducibly complex" and could not, therefore, have been built up gradually in little steps in a Darwinian manner.[7]

At the pinnacle of biological complexity is the human brain, the most complex structure known to exist in the universe. It contains about 100 billion neurons, each of which has connections to as many as a thousand other neurons. It dwarfs in sophistication any computer ever devised by man. And yet, this brain, with all its astonishing powers — the brain of Mozart, of Einstein, of Shakespeare — evolved from an ape brain in about 5 million years or less.

Why, then, are so many scientists so sure that natural selection is sufficient? Probably because they see no alternative that does not involve some divine superintendence of affairs, and to admit such a possibility would be, they think, "unscientific." My own view is that it is unscientific to go beyond the evidence. A truly scientific person should keep an open mind on how evolution happened. Many scientific people have. In an essay written in 1959, Werner Heisenberg described the views on this subject of Wolfgang Pauli, one of the century's most brilliant physicists: "Pauli is skeptical of the Darwinian opinion, extremely widespread in modern biology, whereby the evolution of species on earth is supposed to have come about solely according to the laws of physics and chemistry, through chance mutations and their subsequent effects. He feels this scheme is too narrow. . . ."[8] As Heisenberg noted, Pauli had a reputation among his colleagues for being a sharp critic who subjected proposed theories "to unsparing criticism of every obscurity and inexactitude." Unfortunately, when it comes to the questions surrounding how evolution happened, many scientists seem to be quite uncritical, not to say dogmatic, in their attitude.

In spite of very good reasons for withholding judgment on the question whether natural selection is sufficient to explain the intricacy of biological structures, I will from now on assume for the sake of argument that it is.

Does Darwin Give "Design without Design"?

We saw in chapter 11 that marvelous patterns can arise "spontaneously" in nature as a result of the normal operation of the laws of physics. However, we also saw that this could only happen because of even more marvelous patterns in the laws of nature themselves. In other words, behind the remarkable phenomena stood even more remarkable laws.

The same is true of evolution. Evolution, presumably, occurred through the normal operation of natural laws. But this was only possible, as I shall argue in the next chapters, because the laws of nature are themselves quite special. The biologist Richard Dawkins, referring to William Paley's "watch argument," calls the universe the "Blind Watchmaker."[9] The "watches," for Dawkins, are the intricate structures of living things. The universe, mindlessly following its mechanical laws, has succeeded in crafting these astonishing structures by repeated trial and error. What Dawkins does not seem to appreciate is that his Blind Watchmaker is something even more remarkable than Paley's watches. Paley finds a "watch," and asks how such a thing could have come to be there by chance. Dawkins finds an immense automated factory that blindly constructs watches, and feels that he has completely answered Paley's point. But that is absurd. How can a factory that makes watches be less in need of explanation than the watches themselves? Paley, if still alive, would be entitled to ask Dawkins how his Blind Watchmaker came to be there. Perhaps Dawkins would answer that it was produced by a Blind "Blind Watchmaker" Maker.

It is a remarkable thing that inanimate matter assembled itself into living organisms like dogs and cats and chimpanzees. The fact that it happened according to natural processes makes it no less remarkable; on the contrary, it only shows how remarkable the natural processes of our universe are. It is the same with sexual reproduction. Suppose someone told us of a technological breakthrough whereby a microscopic pellet of chemicals could be placed in an appropriate bath of other chemicals and spontaneously assemble itself into a video camera, or pocket calculator, or power tool. Would we not be astonished that such a thing was even possible, let alone that someone had achieved it? And yet, sexual reproduction is a more amazing thing by far. A little pellet of chemicals assembles itself into organisms that are far more sophisticated than anything human engineers can design or build. Humans can make a jumbo jet or a fighter plane, but are nowhere near to being able to make something as sophisticated as a housefly or a mosquito.

And evolution is a far stranger thing even than reproduction. For what evolution means is that from a soup of very simple particles there emerged spontaneously all the complex entities that are capable of reproducing themselves. The entire genetic *system* by which organisms assemble themselves from microscopic seeds spontaneously assembled *itself* from a mere bath of chemicals.

The fact that nature has the capacity to do these things should arouse wonder and puzzlement. It forces us to confront the question of whether there is something special about the laws of nature themselves that makes it possible. Or would *any* kind of universe, governed by *any* kinds of laws, have the same capacity to bring forth life spontaneously? This is the question we shall examine in the next chapters. What we shall find is that our universe's openness to biological evolution appears to be a consequence of the fact that its laws are indeed very special. A slightly different set of laws would, it seems, have led to a completely lifeless, sterile universe.

If this is so, then Darwinian evolution, far from disproving the necessity of a cosmic designer, may actually point to it. We now have the problem of explaining not merely a butterfly's wing, but a universe that can produce a butterfly's wing.

PART IV

Man's Place in the Cosmos

14 | The Expectations

What is the place of man in the cosmos? This is one of the central issues that divides the materialist and the religious believer. To the Jew or Christian, the universe was created, at least in part, for the sake of the human race. The Epistle to Diognetus, a Christian work of the early second century, says, "God loved the race of men. It was for their sakes that he made the world."[1] In the Book of Genesis, the six days of Creation culminate in the creation of man, who alone among all the creatures of Earth is said to be made "in the image of God." This does not necessarily imply that other intelligent beings do not exist elsewhere in the universe. Traditionally, theologians have explained that human beings are made in God's image primarily because, like God, we have reason and free will. If there are other beings in the universe endowed with rationality and freedom, then they too are made in the image of God and presumably are no less important in the divine plan. In any case, from the religious perspective, it can be said that it was partly for the sake of the existence of rational, free creatures such as ourselves that God created the universe.

In the view of the materialist, however, it makes no sense to talk about the universe existing for anybody's sake, or for any purpose whatsoever. The universe just is. It is a brute fact, without cause and without purpose. In the words of Victor Stenger, author of *Not By Design*, "The simplest hypothesis that so far seems to explain the data is that the universe is an accident."[2] In his best-selling book *The First Three Minutes*, Steven Weinberg, the leading particle physicist of our time, wrote,

> It is almost irresistible for humans to believe that we have some special relation to the universe, that human life is not just a farcical outcome of a chain of accidents, . . . but that we were somehow built in from the beginning. . . . It is very hard for us to realize that [the entire Earth] is just a tiny part of an overwhelmingly hostile universe. . . . The more the universe seems comprehensible, the more it also seems pointless.[3]

115

Not only to Weinberg, but to many scientists, the progress of scientific understanding has more and more made the universe appear "pointless," and the
human race seem to be merely a by-product of blind material forces. Indeed, it
is believed by many that this is the key lesson that science has to teach us. A particularly forthright champion of this view is the biologist Richard Dawkins, who
writes, "The universe we observe has precisely the properties we should expect
if there is at bottom no design, no purpose, no evil, no good, nothing but blind,
pitiless indifference."[4]

The pointlessness of the cosmos and its indifference to human beings is also
a main theme in the writings of the zoologist Stephen Jay Gould, as in this passage of his book *Full House*:

> I have often had occasion to quote Freud's incisive . . . observation that all
> major revolutions in the history of science have as their common theme . . .
> the successive dethronement of human arrogance. . . . Freud mentions three
> such incidents: We once thought that we lived on the central body of a
> limited universe until Copernicus, Galileo and Newton identified the earth
> as a tiny satellite to a marginal star. We then comforted ourselves by imagin-
> ing that God had nevertheless chosen this peripheral location for creating a
> unique organism in His image—until Darwin came along and "relegated
> us to descent from an animal world." We then sought solace in our rational
> minds until, as Freud noted, . . . psychology discovered the unconscious.[5]

Gould goes on to observe that Freud left out "several important revolutions in
the pedestal-smashing mode." One is the discovery of "deep time":

> The earth is billions of years old. . . . The Freudian dethronement occurred
> when paleontology revealed that human existence only fills the last micro-
> moment of planetary time—an inch or two of the cosmic mile, a minute or
> two of the cosmic year.[6]

The human race is, according to Gould, a freak accident of evolutionary history,
merely "a tiny twig on an ancient tree of life."[7]

This idea of the progressive "dethronement" or marginalization of man by
scientific discovery is perhaps the central claim of scientific materialists. It lies
at the core of their view of reality. The question is whether it is justified by a
dispassionate examination of the scientific data, or is based on their own philosophical preconceptions.

It cannot be denied that there is much in the history of scientific discovery
that lends itself to this marginalization-of-man interpretation. And yet, discussions about the size or the age of the universe do not come to grips with

the real question: Is the human race an "accident" or is it a central part of the cosmic plan? Or, to put it in Weinberg's terms, *are* we "somehow built in from the beginning"?

As it happens, some new light has been shed on these old questions by ideas that have come out of physics and astronomy in the last few decades. In particular, starting with the work of the astrophysicist Brandon Carter[8] in the 1970s, it has been noticed that there are quite a few features of the laws of physics that seem to suggest precisely that we *were* built in from the beginning. Some scientists call these features "anthropic coincidences." ("Anthropic" comes from the Greek word *anthropos,* meaning "human being.") What is "coincidental" is that certain characteristics of the laws of physics seem to coincide exactly with what is required for the universe to be able to produce life, including intelligent beings like ourselves.

Many, including former agnostics and atheists, have seen in these anthropic coincidences a powerful argument for the existence of God. Others have argued that these coincidences can be explained scientifically without invoking God. Their discovery has not, in short, succeeded in ending the old debate between religion and materialism. That is hardly surprising. Nevertheless, it has dramatically changed the terms of that debate. It is no longer a question of whether one can find any evidence in nature that we were built in. Such evidence abounds. It is now a question of whether that evidence should be taken at face value, whether it really means what it seems to mean.

In the next chapter I will present and explain some examples of anthropic coincidences. In subsequent chapters I will discuss their implications.

15 The Anthropic Coincidences

It has been found that many features of the laws of physics seem to coincide exactly with what is required for the emergence of life to be possible. This is what is meant by the term *anthropic coincidences*. How does one go about deciding if a certain feature of the laws of physics is indeed an anthropic coincidence? The obvious answer is that one analyzes what the universe would have been like if that feature had been different. In other words, one takes the equations that describe our universe, makes an alteration in them, and then attempts to figure out what the hypothetical universe described by those new equations would have been like. For example, one could ask, "What if the mass of the electron had not been 0.511 MeV, but some other number? How would the universe have been different? Could life have evolved in it?"

Answering such questions in a complete way is undoubtedly beyond the ability of human beings. A different electron mass, for example, would have ramifications too numerous and complex to analyze in full detail. It would make a difference in how the elements were made in the early universe, how stars burn, how planets form, and how chemistry works. Nevertheless, anthropic questions are not hopeless. There are some cases where "changing" a feature of the laws of physics (which we can do, of course, only in thought) would have effects so drastic as to almost certainly preclude life. In other words, there are some fairly solid and convincing examples of anthropic coincidences.

In this chapter, I will present eleven examples that I think are fairly sound. I present so many to drive home the point that one is not talking about one or two flukes, but that a large number of such coincidences exist. I could easily have discussed more.[1] As it is, I may have put in more examples than the average reader will want to go through. The reader with a smaller appetite for physics will lose nothing essential to the later discussions if he or she skips over a few of the examples.

118

The first two examples will take the most time to explain, but they are important, as they are among the most famous and frequently discussed examples of anthropic coincidences.

Example 1: The Strength of the Strong Nuclear Force

A very well known argument[2] says that if the strong nuclear force were even slightly weaker or stronger than it is it would have been disastrous for the possibility of life. The strong nuclear force (also known as "the strong interactions") is one of the four basic forces in nature. The other three are gravity, electromagnetism, and the weak interactions.

The question, then, is this: What difference would it have made, especially for the possibility of life, had the strong force been a bit weaker or a bit stronger, across the board, than is actually the case? Before answering that, one must know what the strong force does. One of the important things it does is hold atomic nuclei together. So changing the overall strength of the strong force would change the way atomic nuclei are held together. This would change what kinds of atomic nuclei could exist in nature, and consequently what kinds of atoms could exist. Obviously, this would have profound effects on everything made of matter.

To discuss this, we need to know a few very basic things about atomic nuclei. These will take just one paragraph to explain. Every kind of atom, i.e., every chemical element, has its own kind of nucleus, made up of a specific number of protons together with some number of neutrons, which is different for different "isotopes" of the element. For example, what makes a carbon atom a carbon atom is that it has exactly six protons in its nucleus; however, different isotopes of carbon have different numbers of neutrons. Thus, one isotope of carbon has six protons and six neutrons in its nucleus, for a total of twelve, and is therefore called "carbon 12." Another isotope has six protons and eight neutrons, for a total of fourteen, and is called "carbon 14"; and so on.

There are almost one hundred naturally occurring chemical elements. The smallest is hydrogen, with one proton in its nucleus, the largest is uranium with ninety-two protons. Twenty-five of these elements are found in the human body and seem to be necessary for its functioning.[3]

Where did all these elements come from? Were they all there at the very beginning of the universe? The answer is no. They were manufactured in three places: in the fires of the Big Bang, in the interiors of stars, and in the explosions of stars called "supernovas," which also have the effect of spewing the elements made in stars out into space, where they can be used to make planets, plants, and people. You and I are made of stuff that was once in the deep interior of a star. Some physicists like to joke that we are made of stardust—and it is true.

The way that elements are made is that first their nuclei are made by the fusing together of smaller nuclei. Then, later, the nuclei get clothed with a cloud of electrons to make a complete atom. Since atomic nuclei get made from the fusing together of smaller nuclei, the whole process must have started off with the smallest possible nuclei. To begin with, the only nuclei consisted of just single particles: either lone neutrons or lone protons. (A lone proton is the nucleus of the simplest form of hydrogen: hydrogen 1.)

The next step is obvious. Pairs of particles had to fuse together to make two-particle nuclei, and that is what happened. Specifically, lone protons fused with lone neutrons to make hydrogen 2 (usually called deuterium," from the Greek word *deuteros* meaning "second"—it is the second thing to form). From there, two steps were possible: adding a single-particle nucleus to a two-particle nucleus to make a three-particle nucleus, or adding together a two-particle nucleus and another two-particle nucleus to make a four-particle nucleus. (1 + 2 = 3 and 2 + 2 = 4. Not everything in nuclear physics is difficult.)

And so it went, with smaller nuclei fusing into larger ones, until all the elements up to uranium were made. Only the first few steps happened in the Big Bang. The heavier elements required cooking inside stars and supernovas.

Now back to the question with which we started: What difference would the strong force having been a little weaker have made? The answer is that making the strong force only about 10 percent weaker would choke off the process of making the elements at the very beginning, at the very first step.

The first step, we just saw, was putting one proton and one neutron together to make the two-particle nucleus called hydrogen 2 or deuterium. But if the strong force were about 10 percent weaker, it would not be strong enough to hold a deuterium nucleus together. Even in the world as it is, deuterium is just barely held together by the strong force. (It takes only about one-fourth as much energy to rip it apart as it does to rip a particle out of a more typical nucleus.) If the strong force were just 10 percent weaker it could not do the job at all. Protons and neutrons would bounce off each other without sticking.

That, of course, would be a disaster. Because without that first step in the chain, none of the subsequent steps could occur. No elements would form except the very first: hydrogen 1. Instead of nearly one-hundred elements in nature, only hydrogen would exist, and one cannot make a living thing out of just hydrogen. If that were not bad enough, there would be no sunlight to be the energy source that life needs. The Sun and other stars burn by nuclear fusion reactions—that is, by those processes in which little nuclei get fused together to make bigger ones. With particles unable to stick together to make deuterium, fusion would not get off the ground. With no Sun or similar stars, and with nothing around to make things out of except gaseous or liquid hydrogen, the chances of life would seem to be utterly negligible or zero.

What about the opposite scenario? What if the strong force were only a little bit *stronger* than it is? In that case, there would also be dramatic, and possibly catastrophic, consequences. It has been estimated that if the strong force were only 4 percent stronger than it is new kinds of two-particle nuclei would be possible: two protons could stick together to make a "di-proton," and two neutrons could stick together to make a "di-neutron." (In our world these two nuclei cannot exist.) This would have dramatic consequences for how stars burn. In our world, the first step in the burning of hydrogen in stars like the Sun has two protons getting together to form deuterium, a process that requires that one of the protons convert into a neutron through a "weak interaction." This involvement of the "weak interaction" makes the burning proceed very slowly and is the reason that stars like the Sun last for billions of years. However, in a world with a stronger strong force hydrogen would be able to burn in stars directly by fusing into di-protons, a process that does not require a weak interaction and is therefore very fast. Stars would last for so short a time that life, it is argued, would not have time to evolve.

Example 2: The Three-Alpha Process[4]

In the first example we traced the building up of the elements to the point where nuclei made up of four particles appeared. The only kind of nucleus with four particles that can exist is helium 4, which has two protons and two neutrons.

Making helium 4 is very easy; in fact a great deal of helium 4 was made in the Big Bang and still is made in stars. However, the next step in the chain is exceedingly tricky. The problem stems from the fact that four particles prefer to be bound together as a helium 4 nucleus. Helium 4 nuclei even have a special name of their own—"alpha particles"—for historical reasons that we need not go into. One could say that the particles in a helium 4 nucleus form a kind of exclusive club. If a fifth particle comes along and tries to join the club, it is rejected. Similarly, if two helium 4 nuclei try to get together to form a larger nucleus, they fail: the eight particles would rather remain as two helium 4 nuclei than as one larger nucleus. That is why there are no stable nuclei that contain five or eight particles. These are the famous "gaps" at five and eight. What this means is that once helium 4 forms, it is hard to make anything larger. In other words, after making helium 4 the next rungs of the ladder seem to be missing.

When scientists first started thinking about how the elements formed (nucleosynthesis), they had a hard time understanding how any of the elements in nature larger than helium 4 got produced. How did nature accomplish the feat? The way it happened (and still happens) is rather delicate. In the interiors of stars, helium 4 nuclei are constantly bouncing into each other but are unable to stick together. However, they do remain together for a very brief interval of time: about a hundred-millionth of a billionth of a second (10^{-17} sec). It sometimes happens that just in that fleeting moment when two heium 4 nuclei are joined a third helium 4 comes along and hits them. And *three* helium 4 nuclei, as it turns out, *can* stick together to form a stable nucleus. In fact, the nucleus they form is just carbon 12. Nature, in other words, takes a large "double step" to get past the missing rungs in the nucleosynthesis ladder. Once carbon nuclei are formed, it is clear climbing to make all the heavier elements. There are no more missing rungs.

Since helium 4 nuclei are also called alpha particles, this curious process where three of them collide at once is called the "three-alpha process." As one can imagine, the three-alpha process does not happen very often, since the third alpha has to come along at just the right moment—just in the right hundred-millionth of a billionth of a second. In fact, when physicists first estimated how much carbon would be produced in stars in this way, they found much less than actually exists in nature. This was baffling.

The solution to the puzzle was suggested by Fred Hoyle. (We came across his name earlier, as one of the founders of the Steady State Theory of cosmology.) Hoyle suggested that the three-alpha process was "resonantly enhanced." The idea of resonant enhancement can be illustrated by some familiar examples from everyday life. We have all experienced the windows of a house suddenly starting to rattle when a heavy truck goes by or when there is some other loud and steady noise. The reason this happens is that a window has certain "normal modes of vibration"—that is, certain ways it naturally rattles. If the sound waves from a truck happen to very closely match one of a window's normal modes of vibration, the window will react strongly.

The same phenomenon happens in many situations. Singing in the shower, one sometimes hits a note that resonates strongly—the sound is unexpectedly loud and full. That is because one has struck one of the normal modes of vibration of the air in that room. Or if one blows across an open bottle top in the right way, a loud, resonant sound is produced. Some people even claim that an opera singer, by hitting the right note, can shatter a wine glass.

The same kind of thing happens in nuclear physics. Nuclei also have certain normal modes of vibration. If nuclei come together to form a new nucleus and have just the right amount of energy, an amount that happens to match one of the normal modes of vibration of the new nucleus, then the fusion reaction can be strongly enhanced. What Hoyle suggested is that carbon 12 nuclei must happen to have a normal mode of vibration—in physics jargon, an "energy level"—which is at just the right energy to enhance the three-alpha process. Only if there were such an enhancement, he reasoned, could enough carbon have been produced in the interior of stars. Hoyle was able to calculate that the energy of the required energy level of carbon 12 had to be about 7.7 MeV for the three-alpha process to be enhanced. At that time, the energy levels of carbon 12 were not known very well. But, in response to Hoyle's prediction, nuclear experimentalists undertook to determine what they were, and they discovered that indeed carbon does have an energy level at 7.66 MeV.

What if this energy level of carbon had been at a slightly different energy? What if it had been at 7.5 MeV or 7.9 MeV instead? In that case the three-alpha process would not have been resonantly enhanced, very little carbon would have been synthesized in stars, the building up of the elements would have been stymied, and there would be very little ordinary matter in the universe except hydrogen and helium. With only those two elements, almost no chemical reactions would have been possible. In fact, helium is chemically inert, so we would be no better off than we were in the case we discussed before where the only element was hydrogen. Life based on chemistry would have been impossible.

In other words, if the energy levels of carbon had been different by only a few percent, in one direction or the other, carbon-based life (including human beings) would not have been able to exist.

This is not the only fortuitous circumstance that Hoyle noticed. He thought about the next steps in the ladder after carbon is produced. One of these steps is a reaction whereby a helium 4 nucleus combines with a carbon 12 nucleus to make a nucleus with sixteen particles: oxygen 16. If *this* reaction had also been resonantly enhanced, it would probably have been disastrous, since virtually all the carbon would have been "burned up" in stars to produce oxygen 16. No carbon, no carbon-based life. As it happens, the reaction of helium 4 and carbon 12 to form oxygen 16 just barely misses being resonantly enhanced. There is an energy level in oxygen 16 at 7.1187 MeV; if this energy were larger by only a few percent, there would have been a problem.

Example 3: The Stability of the Proton

Not all types of particles are stable; many of them "decay" or disintegrate after a while into other types of particles. The "half-life" tells how long this typically takes. For example, neutrons have a half-life of about ten minutes. Neutrons usually disintegrate into a proton, an electron, and an anti-neutrino.

Fortunately, protons are stable. (At least, it is known that they last much longer than the age of the universe.) That is fortunate because the nucleus of ordinary hydrogen (hydrogen 1) consists of just a proton, and if that were unstable, there would be no ordinary hydrogen in the world. And without hydrogen, there would be no water, no organic molecules, no hydrogen-burning stars like the Sun—in short no possibility of life as we know it.

Why isn't the instability of the neutron equally disastrous, considering that all nuclei except hydrogen 1 contain neutrons? And, for that matter, how can it be that there are neutrons in all those nuclei, like carbon 12 and oxygen 16, if neutrons do last only ten minutes or so? The answer is that whereas an isolated neutron is unstable, a neutron can be quite stable if it is inside an atomic nucleus with other neutrons and protons. The reason for this has to do with a subtle quantum effect called "Fermi energy." The fact that isolated neutrons are unstable does not matter very much, since isolated neutrons, having no electric charge, do not bind together with electrons to form atoms.

The interesting question is how the universe avoids the disaster—not for it but for us—of *protons* being unstable. Why is the proton stable and the neutron unstable? The key is that the neutron is a tiny bit heavier than the proton. The mass of a neutron is 939.565 MeV, while the mass of a proton is 938.272 MeV—a difference of only a seventh of a percent. Relativity theory tells us that mass is the same thing as energy, so this is the same as saying that a neutron has a little bit more energy packed inside it than a proton does. Because of that, a neutron can decay into a proton plus some other particles, and release energy in the process. But a proton cannot decay into a neutron because it does not have enough energy to do so. Because a neutron has a little more energy inside it than a proton, extra energy would have to be supplied to a proton to get it to turn into a neutron.

If things were the other way, if the proton's mass were even a fraction of a percent *larger* than the neutron's, then neutrons would be stable and protons would be unstable, which means that there would be no hydrogen 1, and we would not be here.

The reason for the happy fact that protons are slightly lighter than neutrons has to do with the properties of the quarks out of which protons and neutrons are made. The *u* quark is slightly lighter than the *d* quark, and a proton has a preponderance of *u* quarks in it, while the neutron has a preponderance of *d* quarks. However, nobody yet knows why *u* quarks are lighter than *d* quarks rather than the other way around.

(I should clarify a few points here for the experts, in order to avoid possible misunderstandings. If protons were heavier than neutrons, then isolated protons would not be stable, and the universe would contain no hydrogen 1 atoms. The universe would therefore lack ordinary water and most organic molecules, and there would be no hydrogen-burning stars. However, elements heavier than hydrogen would exist, and there would be stars that burned those other elements. This would happen as follows. Even though isolated protons would be unstable, there would be many of them around right after the Big Bang at a time so early that they hadn't had time to decay. Most of these primordial protons would be fused into helium, where they would be stabilized by Fermi energy. This helium would eventually condense into helium stars, which would, in the process of burning, produce yet heavier elements. Such a universe would possibly have a very rich chemistry, and would certainly have stars. But, since it would lack hydrogen 1, none of the organic chemistry that takes place in our universe would be possible. It seems highly doubtful that life would be possible.)

Example 4: The Strength of the Electromagnetic Force

The intrinsic strength of the electromagnetic force is controlled by a parameter in the laws of physics called "the fine structure constant." The value of the fine structure constant is close to $\frac{1}{137}$. The corresponding parameter controlling the strength of the strong nuclear force is about equal to 1. In other words, the electromagnetic force is intrinsically about one-hundred times weaker than the strong nuclear force. For example, if two protons are close together, the strong-force attraction between them is about one-hundred times greater than the electrical repulsion.

This fact has many important consequences. In particular, it is a crucial part of the explanation of the fact that there are about one-hundred kinds of chemical elements. If the electromagnetic force were of different strength it would change the number of elements that could exist in nature, and therefore the kind of chemistry that could take place. If electromagnetism were significantly stronger it would make it impossible for life as we know it to exist.

Let us see why the strength of the electromagnetic force matters for how many elements exist. The protons inside an atomic nucleus exert an electrical repulsion on each other. The only reason that nuclei do not fly apart is that the protons and neutrons are held together by the more powerful—but shorter-range—strong nuclear force. However, for a very large nucleus, an individual proton only feels the strong-nuclear-force attraction of those protons and neutrons which are nearby to it, whereas it feels the electrical repulsion of *all* the other protons in the nucleus. What this means is that if the nucleus has a large enough number of protons in it, the electrical repulsion of the protons will overpower the nuclear force that is holding the nucleus together, and blow the nucleus apart. In other words, the intrinsic weakness of the electromagnetic force between any two protons can be more than made up for by the sheer numbers of protons in a nucleus. Consequently, it is impossible for there to be stable nuclei containing more than a certain number of protons. That is why there is a limit to the number of chemical elements in nature.

What is that limit? A real calculation of it is quite difficult. But one can make a very crude estimate based on the relative strengths of the electromagnetic force that is trying to blow the nucleus apart and the strong nuclear force that is trying to hold it together. The electromagnetic force is about one hundred times weaker than the nuclear force, so one might guess that when there get to be somewhere on the order of a hundred protons in the nucleus, the electric repulsion wins, and the nucleus disintegrates. Actually, crude as the reasoning is, it is basically correct, and the answer it gives is not too far off. The largest nucleus that occurs in nature—uranium—has ninety-two protons.

What if the electromagnetic interactions were a little stronger? What if the fine structure constant were not $\frac{1}{137}$ but, say, $\frac{1}{25}$? One might expect then that there would be about twenty chemical elements in nature. That would mean that less than half of the twenty-five elements that make up the human body would exist. In particular, there would be no potassium, calcium, or iron, to mention only a few of the elements most important for life. If the fine structure constant were $\frac{1}{10}$, then only a handful of elements would exist, probably not including such basic ingredients of biochemistry as nitrogen and oxygen.

I have only considered here one effect of a stronger electromagnetic force. But there would be other effects that would be even more unfavorable for the evolution of life. For example, if the fine structure constant were only twice as large as it is, then protons would be heavier than neutrons rather than the other way around, with all the baneful consequences mentioned in example 3: no ordinary hydrogen, no water, no organic chemistry, and no hydrogen-burning stars like the Sun. The reason that the proton would be heavier is simple. Protons have a net electrical charge, whereas neutrons do not. That means that protons have a little bit extra electrical energy packed into them that tends to make them heavier than neutrons. Fortunately for us, this is more than compensated for by the fact that the quarks in a neutron are a slight bit heavier than those in a proton. However, if the intrinsic strength of the electromagnetic force were made larger by a factor of two or so, the electrical effect would be more important than the quark-mass effect, and protons would indeed be heavier than neutrons.

Example 5: The Value of v

A critically important number in the present theory of particle physics is a parameter called v. It has to do with the dynamics of a field called the Higgs field. The long technical name of the parameter v is "the vacuum expectation value of the Higgs field." For our purposes the only thing that it is important to know about v is that it determines the masses of almost all the fundamental particles in nature. For example, the mass of the electron is proportional to v. So are the masses of all the quarks. (Recall that protons and neutrons contain quarks, as do all the other particles that feel the strong nuclear force.) There are particles called W^+, W^-, and Z^0 whose masses are also proportional to v. These three particles "mediate" the weak nuclear force, i.e., the weak nuclear force arises from W^+, W^-, and Z^0 particles being exchanged back and forth between other particles. The masses of "neutrino" particles probably also depend on the value of v, most likely being proportional to the square of v, but this is still uncertain.

In any event, a change of v would have profound and far-reaching implications for physics. In some recent papers it has been argued that even a very slight change in v, in one direction or the other, would greatly diminish or completely destroy the prospects for life in the universe.[5]

The parameter v is an energy. To make the discussion simpler, let us agree to call the value that v has in the real world 1. If I talk about v being 5, then I mean that it is five times its real-world value.

The value of v is a great puzzle to particle theorists; in fact, it is one of the central puzzles in physics. What is puzzling is that in reasonably simple theories v seems to want to come out to be, not 1, but a number like 10^{17}, i.e., 100,000,000,000,000,000. The reason for this is that there are various physical effects which are known to contribute that much to the value of v; and for v to come out instead to be 1 means that all of these huge contributions must somehow cancel each other to one part in 100,000,000,000,000,000. It is as though there were a bank account in which many deposits and withdrawals were made in eighteen-figure amounts of dollars, but the balance in the account after all this activity came out to be just one dollar. One would feel that there had to be a reason why those deposits and withdrawals balanced in just that way. But in physics no one knows why the actual value of v is 1 when it seems more natural for it to be something like 100,000,000,000,000,000.

As far as the possibility of life emerging in our universe is concerned, it would be a disaster for v to be 100,000,000,000,000,000. It would also be a disaster if it were 100,000,000,000,000, or if it were 100,000,000,000, or if it were 100,000,000, or if it were 100,000, or even if it were 100. Indeed, it would be a disaster if it were 10, or 5, or even 1.5. It would probably be a disaster if v were even slightly different from the value it happens to have in the real world. Let us see why.

Suppose v were greater than about 1.4. Then deuterium could not exist. We have already seen some of the catastrophic consequences of that in example 1. It would be difficult if not impossible for any elements other than hydrogen to be produced.

The reason that deuterium could not exist if v were greater than about 1.4 has to do with the fact that the masses of fundamental particles depend on the value of v. In particular, if v were 1.4, then the masses of the quarks would be 1.4 times larger than their real-world values. That, in turn, would make a particle called the "pion," which is made up of quarks, have a mass about 1.2 times its real-world value. The pion plays a crucial role in "mediating" the strong nuclear force felt by protons and neutrons. So, if v were 1.4, the "range" of the strong nuclear force would be shorter by about 20 percent than in the real world. This would be enough to weaken the binding of the deuterium to the point where it would no longer hold together. That would choke off the production of elements heavier than hydrogen.

If v were larger still, other nuclei besides deuterium would be unstable as well. This is partly because the range of the strong nuclear force would shrink still further, but also because the neutrons inside nuclei would become increasingly unstable. As we saw in example 3, lone neutrons are unstable because they are slightly heavier than protons; and they are slightly heavier because of the masses of the quarks out of which they are made. If v were larger, then the masses of the quarks would also be larger, and the difference in mass between the neutron and proton would be increased. That would increase the tendency of neutrons to decay into protons, even inside nuclei.

If v were larger than 5, it can be shown that neutrons would be so unstable that no nuclei containing neutrons could exist at all. But since the only kind of nucleus in nature that has no neutrons is ordinary hydrogen, this would lead to a universe with no elements except hydrogen. We have already seen how bad that would be for the prospects for life.

If v were greater than about 500, things would take an even more radical turn. Protons and neutrons are made up predominantly of two kinds of quarks, called u and d. If v were bigger than 500, all d quarks would decay into u quarks, meaning that there could be no neutrons or protons. In such a universe there would be instead particles similar to protons and neutrons but containing only u quarks. Such particles are called Δ^{++} and have twice the electrical charge of protons. They would be unable to fuse together to make larger nuclei, so the only nuclei that would exist would consist of a single Δ^{++}. When clothed with electrons these would form atoms chemically identical to helium, which is an inert element. Since all matter would be an exotic form of helium, there would be no chemistry and the universe would be completely sterile. For all values of v between about 500 and the "natural value" of about 100,000,000,000,000,000, this is the kind of universe that would exist.

One can also ask what would happen if v were much *smaller* than it is in the real world. It turns out that this would probably have been just as disastrous for the chances of life, though in very different ways. These are a little more complicated to explain, so I will not go into them.

The astonishing fact seems to be that, of all the huge range of possible values of this parameter v, it must fall within a very narrow range if there is to be any chance of life evolving, at least life that is made up of atoms and sustained by energy from stars.

The papers in which these conclusions were arrived at were probably, if anything, too conservative. They argued only that v had to be within about 40 percent of its real-world value if disasters such as unstable deuterium were to be avoided. But probably even much smaller changes of v away from its real-world value would spell disaster. For example, it is likely that a change of v by only a few percent would shift the energy levels of carbon by enough to wreak havoc with the production of carbon by making the three-alpha process non-resonant.[6]

(See example 2.) It seems probable that v must be extremely precisely chosen to allow the evolution of life to occur in a way at all resembling the manner that it did.

Example 6: The Cosmological Constant

In the equations that govern the gravitational force (Einstein's equations), there are two numbers. One of them is called Newton's constant, and is conventionally represented by the symbol G_N. It says how strong gravity is. The other number is called the cosmological constant,[7] and is conventionally represented by the Greek letter lambda (Λ). The cosmological constant tells how much gravitational pull is exerted by "empty space." (This may sound absurd, but in quantum theory empty space is not as empty as it seems—it seethes and bubbles with "quantum fluctuations.") As we saw in the chapters on the Big Bang, the value of the cosmological constant is important to how the universe as a whole behaves.

In discussing the size of the cosmological constant we shall use what are called "natural units" for gravity, in which Newton's constant is exactly 1. It has long been known that the cosmological constant (when expressed in these natural units) is less than about 10^{-120}. In decimal form this would be written 0.0000000000000000000000000000000000000000000000000000000000000 0000000000000000000000000000000000000000000000000000000001. This is an amazingly small number. It is so small that physicists have long assumed that the cosmological constant is really exactly zero. But it is hard to tell whether some physical quantity is exactly zero, or just too small to be measured with available techniques. Recently, there have been astrophysical measurements that seem to imply that the cosmological constant is not exactly zero, but rather is a number about 10^{-120}.

In either case, whether the cosmological constant is exactly zero or just fantastically small, physicists are confronted by a very deep puzzle. In physics, if a number is either exactly zero or extremely small there is usually a physical reason for it. For example, the mass of the photon is believed to be exactly zero; that is understood to be the consequence of a fundamental symmetry of the laws of physics called "electromagnetic gauge invariance." So far, no one has been able to find the physical reason why the cosmological constant is small or zero. This failure is the so-called "cosmological constant problem," and is considered by many scientists to be the deepest unsolved problem in physics. (The smallness of v compared to its "natural value" is another puzzle of the same kind.)

It turns out to be a very fortunate thing for us that the cosmological constant *is* so small. If it were not, the universe would not have been able to have a nice steady existence for the billions of years required for life to evolve.

One has to consider two cases, because the cosmological constant is allowed by the mathematics to be either a positive or a negative quantity. Suppose, first, that the cosmological constant had been negative and equal to some number that was not particularly small or particularly large, say -1. Then the universe would have gone through its entire life cycle of expansion and collapse in the incredibly short time of 10^{-43} seconds. That is, the universe would only have lasted a ten-millionth of a billionth of a billionth of a billionth of a billionth of a second. This very short time is called the "Planck time," and is a fundamental length of time in physics. In essence, the Planck time is the shortest period of time that any physical process can happen in and still be described in terms of our usual notions of space and time.

If we suppose instead that the cosmological constant had been negative and equal to about minus one-millionth (i.e., -0.000001), then the universe would have lasted for a thousand Planck times, namely about a ten-*thousandth* of a billionth of a billionth of a billionth of a billionth of a second. Not a great improvement from our point of view. If the cosmological constant is negative, and the universe is to last for the several billion years required for life to appear, then the magnitude of the cosmological constant has to be less than about 10^{-120}, the terrifically small number mentioned earlier.

Now suppose that the cosmological constant is a positive number. If it had been positive and not too large or small, say about +1, then the universe would have undergone an incredible "exponential" expansion, in which it doubled in size every ten-millionth of a billionth of a billionth of a billionth of a billionth of a second—that is, in every Planck time. The universe would have lasted forever, but always expanding ferociously. Even if the cosmological constant had been as small as $+10^{-48}$, the universe would have expanded so fast that it would have doubled in size in the time it takes an electron in an atom to orbit the atomic nucleus once. In such a situation, even atoms would be ripped apart by the expansion of the universe. Even if the cosmological constant had the much smaller value $+10^{-80}$, the universe would have doubled in size every thousandth of a second or so, which would be so fast that your body would be ripped apart by the expansion. If the universe was to have a sufficiently gradual expansion over billions of years to allow life to evolve, then the cosmological constant had to be less than or about $+10^{-120}$.

In order for life to be possible, then, it appears that the cosmological constant, whether it is positive or negative, must be extremely close to zero—in fact, it must be zero to at least 120 decimal places. This is one of the most precise fine-tunings in all of physics.

Example 7: The Flatness of Space

We have just seen that the universe would not have lasted for billions of years, as was needed for the evolution of life, had not an extremely stringent condition been satisfied by the cosmological constant. But there is another apparent "anthropic coincidence" that also had to occur if the universe was to last so long, namely that space had to be extremely flat. (By flat here I do not mean two-dimensional. I mean lacking curvature — not warped.)

Einstein's theory of gravity tells us that the presence of matter warps space, or more precisely space-time. But if we ignore the local bumps and wrinkles produced by stars, galaxies, and other clumps of matter, and look at the shape of the universe as a whole (or at least the part that can be seen with telescopes), space appears to be extremely flat on the average. Or, to put it another way, the average "spatial curvature" is extremely small.

The amount of spatial curvature of the universe is not, as far as we now know, a fundamental parameter in the laws of physics themselves. Rather, in the Standard Cosmological Model, it enters the equations through the "initial conditions" (or "boundary conditions") of the universe. That is, the spatial curvature is a fact about the universe, rather than a feature of its basic laws. The spatial curvature of the universe changes with time (unless it happens to be exactly zero, in which case it remains zero), and so one has to talk about how big it is or was at a particular time. Which time one chooses is arbitrary, so to keep things simple I will choose one second after the Big Bang.

Let us suppose that the spatial curvature is a positive quantity,[8] and also that the cosmological constant is less than 10^{-120}, as I argued in example 6 that it has to be. Then it turns out that for the universe to have been able to last for the time that it has — about 15 billion years — the spatial curvature one second after the Big Bang had to be less than about 10^{-35}, that is, less than one hundred-millionth of a billionth of a billionth of a billionth. If it had not been, the universe would have expanded and collapsed long before now. Similarly stringent constraints apply if the spatial curvature is a negative quantity.

Example 8: *The Number of Dimensions of Space*[9]

So far, I have given examples based on what might be called the "fine-tuning" of numbers that appear in the laws of physics or in the boundary conditions of the universe. But one can also ask what would have happened had certain grosser features of the laws of nature been different. One example is the number of dimensions of space. We live in a three-dimensional world; what if the world had been two-dimensional or five-dimensional instead?

It may sound silly to talk about the number of dimensions of space as though it could have been any different. We are used to taking the number of dimensions of space for granted. How could it possibly be anything other than three? We cannot even picture what a world with more than three dimensions would be like.

As crazy as it might seem at first, however, physicists take very seriously the idea of universes with more or less than three dimensions of space. Even though such universes cannot be visualized, they can be studied mathematically. In fact, theoretical physicists have devoted an enormous amount of research to them. Partly, this is for the mathematical insights that such hypothetical universes can give into the structure of the real world. For example, often effects that are difficult to analyze in a realistic three-dimensional context can be understood more easily in the mathematically simpler context of one or two dimensions.

But universes with different numbers of space dimensions are studied for another reason as well: namely, it may well be that the real world actually has more than three dimensions. This idea was first proposed seriously by the Polish physicist Theodor Kaluza in 1921. He pointed out that by assuming that the world had a hidden extra dimension of space it was possible to give a unified description of electromagnetism and gravity as a single force.

Interest in theories with "extra dimensions" then faded for several decades. However, it had a resurgence in the early 1980s and has continued strongly ever since. In fact, the theories that are presently regarded as the most promising candidates for the ultimate unified description of all physical phenomena suggest that the universe in which we live may actually have as many as six or seven extra, hidden dimensions of space.

If there are extra, hidden dimensions in our universe they could be hidden from us in two ways. One way is that there are physical reasons that prevent us from moving off or seeing off into the extra directions in space. We are trapped within three of the space dimensions. (As an analogy, consider a truly animated—i.e., living—cartoon figure existing on a two-dimensional surface or sheet, unable to move off or see off into the third dimension, of which he is therefore entirely unaware.) Until very recently, this was *not* the way physicists imagined that there could be extra dimensions. (Though as long ago as 1896 H. G. Wells wrote a brilliant science fiction story, called "The Plattner Story," based on this idea.) However, in the last few years this kind of scenario has generated enormous interest among particle physicists.

The second way that extra dimensions could be hidden from us is that they could be rolled up—in the jargon, "compactified"—to a very tiny size. This can also be understood with a simple analogy. A sheet of paper is two-dimensional. However, one can take a narrow strip of paper and roll it up into a tube—like a paper soda-straw. The soda-straw is still a two-dimensional surface. This would be obvious to us if we were the size of tiny insects. We would then be able to crawl around on the straw in two directions: both along its length and around its circumference. On the other hand, if we were giants much larger than the straw, neither our eyes nor our clumsy fingers would be able to make out the fact that the straw had any thickness. It would seem to us just like a one-dimensional line. From the point of view of the giant, one of the two dimensions is "hidden."

According to speculations of particle physicists, the real universe might have three "large dimensions" of space (analogous to the dimension that goes along the straw) and several hidden dimensions (analogous to the tiny dimension that goes around the straw). Something much smaller than a subatomic particle would experience the world as, say, ten-dimensional, while we gross humans, who cannot see such small distances, are aware only of the existence of the three large dimensions.

Whether or not these new theories turn out to be correct, they have made physicists aware that the number of dimensions cannot be taken for granted. There is no law of logic or mathematics or metaphysics that says that the number of space dimensions has to be three. So it is quite reasonable to ask the question of what the universe would have been like had there been more than three space dimensions or less than three. (In the discussions that follow, when I refer to the number of space dimensions I will always mean the number of "large," macroscopically observable dimensions in which we are able to move.)

If the number of space dimensions had been *less* than three there would have been an obvious problem for any animal that had a brain or nervous system, namely wiring. If one tries to draw the diagram of an electrical circuit in two dimensions, one finds, except for very simple circuits, that the lines representing the wires intersect each other. If the world were two-dimensional it would be impossible to make complex circuitry without wires crossing other wires, thus creating short-circuits. In three or more dimensions, on the other hand, wires (or nerve fibers) can simply go over, under, or around each other, just as a road can go across another road without intersecting it by means of an underpass or overpass. This suggests that any organism with a complex brain and nervous system would have to live in at least three space dimensions.

Other difficulties might have arisen if the world had *more* than three space dimensions. The reason is that the dimensionality of space determines the way that various physical effects depend upon distance. For example, in a three-dimensional world the brightness of a light depends on how far away it is according to an "inverse square law." (If one is twice as far away from a flashlight, it appears one-quarter as bright; if three times as far away, it appears one-ninth as bright; and so on.) However, in a world that had four space dimensions the intensity of light would fall off according an "inverse cube law." And, in general, in a world with N space dimensions, the intensity of light would fall off as the inverse (N-1) power. The reason for the dependence on the number of dimensions is that the falloff in light intensity is due to the spreading out of the beam as it moves away from its source, and this is a matter of geometry: if there are more dimensions of space, then there are more directions in which the light can spread out, and therefore its intensity would fall off more rapidly with its distance from the source.

What is true of the intensity of light is also true of the strength of certain forces. In our world, the electric force and gravitational force also fall off with distance according to an inverse square law. In a world with N space dimensions, however, they too would fall off as the inverse (N-1) power. This would have profound consequences. For example, it would drastically affect planetary orbits. In three space dimensions planets can have stable orbits that are circular, or nearly circular. That is because the pull from the central star can be exactly balanced by a planet's "centrifugal force." However, if the number of space dimensions were larger than three, this balance of gravity and centrifugal force could not exist, and planets would either plunge into the stars they were orbiting or fly off to infinity. Stable orbits would be impossible.

The first person to make this argument was, interestingly, none other than the Anglican clergyman William Paley in his book *Natural Theology*, written in 1802. This is the same William Paley whose famous "watch argument" I discussed in chapter 9.

Similar considerations apply to the orbits of electrons in atoms. In three dimensions, because the electric force obeys an inverse square law, and because of quantum mechanical effects, an electron is able to orbit stably around an atomic nucleus. However, if there were more than three dimensions, then the electrical attraction would cause electrons to plunge into the nucleus and remain there. Atoms would collapse and chemistry, therefore, would be impossible.

I have been considering the number of space dimensions, but the number of time dimensions cannot be taken for granted either. In the real world there is only one time dimension, of course. But it is very easy to write down mathematical theories that describe hypothetical universes with any number of time dimensions. Universes with no time dimension or with more than one time dimension are mathematically quite possible.

A world with no time dimension would, of course, be completely static. It is difficult to imagine what it could possibly mean to have "biology" in such a world. Nor is it any easier to imagine what "life" would be like in a world with several time dimensions. Processes would not flow along in linear progression. Thinking, for example, would not proceed in a sequential manner.

Altogether, then, it seems that for a variety of reasons a universe with three space dimensions and one time dimension is especially likely to be able to support living beings, especially beings having complex processes of thought.

Example 9: The Quantum Nature of the World

Quantum physics is often thought to be some arcane discipline that only has relevance to extremely exotic phenomena. In actuality, it is highly relevant to everyday life. Indeed, in all likelihood, life itself would not have been possible in a non-quantum world.

The quantum nature of the physical world was discovered in the early part of the twentieth century. The way of describing the world that preceded quantum physics is now called "classical" physics. It is important to realize that classical physics was not simply "wrong." Rather, it is the correct way to describe nature in a certain limit or approximation. That is why classical physics still forms the core of physics education. In a great many applications classical physics remains the best way to analyze physical situations. However, if one tries to push classical physics too far, in particular by applying it to atomic and subatomic systems, its inadequacies begin to show up.

Perhaps a rough analogy will help. When one looks at a color picture in a magazine, it appears to be a very accurate image of the object it is supposed to represent. However, if it is examined very minutely, the magazine picture can be seen to be made up of many tiny, colored dots. If one were to examine both the picture and the actual object it represents with a strong magnifying glass, they would look very different. The magazine picture is "grainy," while the real object is continuous or smooth.

In a similar way, the quantum world has a "graininess" that a classical world would not have. However, this graininess is not apparent to a casual observer; in fact, it cannot be seen except with the help of very sophisticated instruments. That is why in many cases the classical description of reality is just as good, practically speaking, as the quantum description.

In the real world, energy comes in little grains called "quanta." For instance, the energy in a beam of light is made up of quanta called "photons." A photon is a particle of light. If the world were governed by classical principles, one could make the energy in a beam of light be as small as one wished. But in the real world, a beam of light cannot have less than one "quantum" of energy. Similarly, the energy in an atom does not vary smoothly, but comes in chunks: atoms have discrete "energy levels."

The graininess of energy in the world saves us from an ultimate catastrophe: the collapse of matter. The point is, if an atom were described by the laws of classical physics, then each electron in the atom would lose energy continuously as it orbited around the nucleus. It would lose this energy by "radiating" electromagnetic waves—that is, light. As it lost energy, it would spiral in closer and closer to the nucleus, until finally it came to rest at the nucleus's center. The whole process would take less than a billionth of a second. Atoms would collapse. Even atomic nuclei would collapse. In fact, all matter would collapse to infinite density.

However, in our quantum world, this collapse cannot happen. An electron in an atom is only permitted to have certain discrete amounts of energy, and the energy it radiates as light can come only in the discrete "photon" packets. This prevents the electron from spiraling down into the nucleus. Rather, its lowest "energy level" is one where it is safely orbiting well outside the nucleus.

Another way of explaining this uses the famous Heisenberg Uncertainty Principle, a central principle of quantum theory, which says that a particle cannot have a definite position and "momentum" at the same time. If the electron really ended up sitting at rest in the center of the atomic nucleus, as it would in a classical world, then it would have both a definite position (namely, the center of the nucleus) and a definite momentum (namely zero, because it would be at rest). However, since the Heisenberg Principle forbids this, the disastrous collapse of the atom does not happen in a quantum world. The Heisenberg Uncertainty Principle is really nothing but a mathematically precise statement of the quantum graininess of the world.

To summarize, the fact that matter, and all the atoms of which it is composed, does not rapidly collapse down to infinite density is due to the fact that the world is governed by quantum principles.

Another consequence of living in a quantum world is that matter is composed of a definite number of types of atoms with stable properties, called "chemical elements." In a classical world, even if the collapse of atoms were somehow prevented, no two atoms would be exactly alike. Given a certain number of electrons sharing a certain amount of energy among them, there would be an infinite number of ways that they could be orbiting around the atomic nucleus. And each electron in an atom could gain or lose any amount of energy depending on circumstances. Atoms would therefore be continuously changing their characteristics as time went on. In a quantum world, by contrast, the electrons can only orbit in fixed ways corresponding to the atomic energy levels. So the properties of an atom are constant in time, and all the atoms of a particular element are exactly the same chemically.

To put it another way, the Periodic Table of the Elements, on which all of chemistry is based, is itself an example of the quantum graininess of nature. Just as a magazine picture is made up of only a certain number of types of dots of color put together in various arrangements, so matter is made up of only a certain number of types of atoms put together in various arrangements.

In a classical world, matter (even if it could be prevented from collapsing altogether) would be endlessly mutating and protean. Any kind of stable structures with predictable behavior would be impossible. Matter would be in constant and unpredictable flux.

A classical world is a logical possibility. In fact, for centuries scientists thought that they lived in one. However, the real world turned out to be governed by the principles of quantum theory, which are notoriously strange and alien to our everyday experience. Had things been otherwise, however, and the world been constructed according to the more "commonsense" principles of classical physics, it is highly doubtful that living structures could have evolved.

Example 10: Why Electromagnetism?

I have already discussed why the strength of the electromagnetic force is crucial to the evolution of life. But an even more basic fact about the world is that there is such a thing as electromagnetism at all. This did not have to be the case. It is quite simple to write down laws of physics that would describe a universe without electromagnetism. In fact, all one has to do is take the equations of our present Standard Model of particle physics and leave out a few "terms" (i.e., pieces of the equations). (Alternatively, one can add a few terms that would have the effect of spoiling or "spontaneously breaking" the basic "gauge symmetries" that are responsible for the existence of electromagnetism.)

What if there had been no electromagnetism? Then there would have been no atoms and no chemistry, since atoms and molecules are held together by electric forces. There would also have been no such thing as light, since light consists of electromagnetic waves. The universe obviously would have been radically different, and it is very hard to imagine living structures having arisen.

Example 11: Why Matter?

Another very basic fact about our universe is that there is such a thing as "matter" at all. Physicists use the word *matter* in various ways. I am using it here, as it is often used in particle physics, to mean "quarks" and "leptons." (Leptons are a kind of particle that includes electrons and neutrinos.) As far as we now know, the Standard Model of particle physics would be just as mathematically self-consistent if there were no quarks and leptons at all.

What would the universe have looked like had there been no quarks and leptons? The answer is that matter would have consisted of "Higgs particles," gravitons, and "gauge particles" (for example, photons, and the W^+, W^-, and Z^0 particles, referred to in example 5). Eventually most of the W^+ and W^- particles would have been annihilated, since the W^+ and W^- are anti-particles of each other, leaving only electrically neutral particles. One cannot, perhaps, absolutely rule out the possibility that some sort of life might have evolved from such unpromising materials, but it is extremely implausible.

16 Objections to the Idea of Anthropic Coincidences

The subject of anthropic coincidences is extremely controversial in the scientific community and provokes a wide spectrum of responses. Many scientists react with hostility or even anger to the mention of these ideas. There are a number of reasons for this.

In the first place, almost all scientists are instinctively and professionally suspicious of anything that smells like "teleology." As the reader will remember from chapter 3, teleological explanations are explanations of things or events in terms of their supposed purposes, goals, or ends. For almost two millennia this kind of thinking prevailed in the physical sciences, and it is generally agreed that it led nowhere. Teleology was found to be a sterile approach to understanding the physical world. Many accounts of the history of science emphasize that the Scientific Revolution occurred only when scientists abandoned teleology in favor of investigating the physical mechanisms that underlie phenomena. That is why any talk about how certain features of the physical world are necessary in order for human life to exist seems to many scientists like a giant step backward, an attempt to smuggle discredited teleological notions back into science. They sincerely worry that people will be led astray from the high road of scientific thinking into the barren wastelands of fruitless metaphysical speculation.

These fears are well illustrated by the following comments that were made by the editor of a major physics journal reacting to a paper on anthropic coincidences that he had received for publication:

> Anthropic arguments of this type are in fact anti-scientific in a very fundamental sense: The essence of the scientific enterprise is the attempt to understand the regularities that we observe in nature; the anthropic argument says that we should not bother. . . . Anthropic arguments can be constructed to

"explain" many phenomena, from the daily appearance of the Sun in the eastern sky, to the equality in magnitude of the electron and proton charges, which most of us now believe to have more conventional scientific explanations.[1]

The basic point that this scientist was making is quite reasonable. One could indeed attempt to explain the fact that the Sun rises every day by saying that if it didn't, we'd never know it—we wouldn't be here. And if people had been content with that as an explanation they would never have attempted to understand how the Sun and planets move, or how the Sun formed and how it generates light and heat. Teleological thinking can indeed be a showstopper as far as doing real scientific research is concerned. For many centuries it was.

Such concerns do not explain all the negative reactions to discussions of anthropic coincidences, however. It is not just the specter of teleology that some scientists fear, but religion. It does not require much imagination to see where thinking about anthropic coincidences can lead.

On the other hand, there are many scientists who disagree that there is anything "unscientific" about investigating anthropic coincidences. In fact, many of the most eminent physicists of our era have taken quite a lively interest in them. The list of those who have written papers on the subject includes such redoubtable names in physics and cosmology as Brandon Carter[2], Andrei Sakharov[3], Bernard J. Carr and Martin Rees[4], Yacov B. Zel'dovich[5], John Barrow[6], Steven Weinberg[7], Lev Okun[8], Andrei Linde[9], Alexander Vilenkin[10], and Stephen Hawking.[11] One of the scientists who has done the most to make anthropic arguments respectable in the scientific community is Steven Weinberg, a Nobel Prize winner, a main architect of the Standard Model of particle physics, and arguably the leading theoretical particle physicist in the world. In recent years he has written a series of papers which point out that the value of the cosmological constant may have an anthropic explanation.

The scientific community, then, is quite divided on the subject of anthropic coincidences. This is illustrated by the case I just referred to, where the editor of a major physics journal criticized a paper on anthropic coincidences as "antiscientific." One of the authors of that paper was, as it happens, himself a former editor of that journal.

I should emphasize that most of the scientists who take an interest in anthropic coincidences do not attribute any religious significance to them. Certainly, for example, Weinberg, who is an atheist, does not. I have already quoted his famous statement that the universe seems "pointless." Most of the authors I have listed take the view that the anthropic coincidences—if they can be explained at all, and are not really *just* coincidences—have a natural scientific explanation in terms of what is sometimes called the Weak Anthropic Principle, an idea that I will explain later. Therefore it is very important to draw the distinction between the "coincidences" themselves and how they are explained or interpreted. There are both religious and non-religious interpretations of them, as we shall see in chapter 17.

In this chapter I am going to deal with some objections that are commonly raised against the very idea that there even are anthropic coincidences at all, or at least that such coincidences have been shown to exist.

THE OBJECTIONS

There are basically three objections to the idea of anthropic coincidences:

1. The Requirements for Life Are Unknown

All anthropic coincidences say that if the universe or the laws of nature had been different in some way, then life would not have been able to evolve. A very basic objection to this is that we cannot really know what is required for life to evolve. The living things we are familiar with have very particular characteristics, which include the following: (*a*) their life processes are based on chemistry; (*b*) their chemistry is critically based on carbon, oxygen, hydrogen, and certain other elements, and on certain critical compounds made from those elements, such as water; and (*c*) the energy that powers life derives ultimately from nuclear reactions taking place inside stars. But how do we know that life of a very different sort could not exist?

For example, it is an old suggestion that life could be based on silicon rather than carbon, or on ammonia rather than water. Some have even suggested that chemistry which uses only the single element hydrogen could be rich enough to lead to living organisms.[12] Life can even be imagined that makes no use of chemistry at all: perhaps the structure of an organism and its life processes could be entirely nuclear rather than chemical. It has even been proposed that life could be based on processes which go on in the hot plasmas in the interior of stars. Alien life might, for all we know, exist inside the Sun. On the other hand, one can imagine life arising even if there were no stars. There are other processes in nature that release energy besides stars, such as gravitational collapse or radioactivity.

To sum up, the supposed anthropic coincidences may all be based on a simple failure of imagination. Who is to say that had the laws of nature been different life could not have arisen in some way that we are unable to conceive?

2. Conventional Scientific Explanations May Exist

A second objection to anthropic coincidences is that the facts which are argued to be necessary for, or at least conducive to, the evolution of life may have conventional scientific explanations. In fact, physicists are already in a position to make good guesses about what some of those explanations are. For example, consider the fourth anthropic coincidence that I discussed in chapter 15, namely the strength of the electromagnetic force, i.e., the value of the fine structure constant. It is tempting to explain the value of this constant of nature by saying that it allows the existence of a variety of chemical elements, and therefore life. However, there is a conventional scientific explanation of the value of the fine structure constant that is based on the idea of the "grand unification" of forces.

Particle physicists have strong reasons for believing that the three non-gravitational forces of nature—the electromagnetic force, the weak nuclear force, and the strong nuclear force—are all parts of a single underlying "unified" force. In fact, there exist very promising theories, called "grand unified theories," in which this is true, and there are even more ambitious theories, called "superstring theories," in which all forces including gravity are unified. A characteristic of such unified theories, not surprisingly, is that the strengths of the different forces are tied to each other. One thing that grand unified theories tend to predict is that the strong nuclear force is much stronger than the weak nuclear force, and that the weak nuclear force is somewhat stronger than the electromagnetic force. In fact, in the simplest grand unified theory,[13] knowing the strength of the weak and strong forces it is possible to calculate what the value of the fine structure constant is. This calculation gives an answer that is correct to within about half a percent. This is one fact that gives some theoretical physicists confidence that grand unified theories are on the right track.

To take another case, the seventh example of an anthropic coincidence that I gave in chapter 15 was the extreme "spatial flatness" of the universe. Here again, scientists think they may have at least a partial explanation of this fact based on the idea of "cosmological inflation." (I have already discussed this idea at some length in chapter 7.)

It is true that most of the facts that I have given as examples of anthropic coincidences are not yet fully explained scientifically. However, the fact that plausible explanations already exist for some of them gives some grounds for hope that most or all of them will eventually be found to have quite conventional scientific explanations.

3. There May Have Been No Room for Choice

The third objection to anthropic coincidences is that one is really not entitled to ask what the universe would be like were some particular feature of the laws of nature different. The reason is that we are not sure which features of the laws could have been different. In fact, there is a sense in which we are not sure that any of them could have been different.

For example, consider the fine structure constant again. In our present theory of particle physics, the Standard Model, the fine structure constant is a "free parameter." In other words, the basic principles of that theory do not require the fine structure constant to have a particular numerical value. In the context of the Standard Model, therefore, it makes perfect sense to ask what would have happened had the fine structure constant had a different value, with everything else in the laws of nature being just as it is. The trouble, however, is that very few physicists think that the Standard Model is the ultimate theory of physics. It is almost certainly only an approximation to some deeper theory, such as a grand unified theory or superstring theory. In such a deeper theory it may not make sense to consider a hypothetical situation in which only the fine structure constant were different. As we just saw, in a grand unified theory the electromagnetic force is tied together with the other forces. Therefore—assuming grand unification is true—had the fine structure constant been different, certain facts about the other forces would have been different also.

In other words, in a deeper theory facts that previously seemed unrelated to each other are often seen to be tied together. Before the principles of electromagnetism were discovered, it would have seemed reasonable to contemplate hypothetical universes where magnetic forces were stronger than in our universe while electric forces were the same. But now that these two forces are understood to be aspects of the same force, it is clear that one of them cannot be different without the other being different also.

The question, then, is whether the kind of reasoning that goes into demonstrating an anthropic coincidence really makes any sense at all. When we know the ultimate mathematical theory of all physical phenomena, it may turn out that everything is so tied to everything else that nothing can be changed without destroying the whole structure of the theory. In the ultimate theory, in other words, it may turn out that everything has to be just as it is. This is what Einstein meant when he famously said, "What I'm really interested in is whether God could have made the world in a different way." Many theoretical physicists believe that in the ultimate theory there will be no freedom: what kinds of particles exist, what kinds of forces there are, how strong those forces are, the number of dimensions of space and time, and all the rest, will be uniquely determined by some powerful and simple set of fundamental principles. If that proves to be the case, then it would be those principles, rather than the requirement that life be able to evolve, that would account for why things are the way they are.

In recent decades many physicists have speculated that it will turn out that the laws of nature are, in some sense, unique: the laws could not have been different than they are. This is a bold idea indeed. What lies behind it?

One thing that lies behind it is the trend in physics toward ever more unified theories: Newton found that the same laws governed earthly and heavenly bodies. In the nineteenth century, electrical, magnetic, and optical phenomena were discovered to be aspects of a single thing called electromagnetism. Quantum theory showed that particles and forces were really two manifestations of the same "quantum fields." Einstein showed that space and time were inseparable parts of a four-dimensional "space-time." In the late 1960s and early 1970s, electromagnetism and the weak interactions were discovered to be parts of a unified "electroweak" force. And there is a great deal of circumstantial evidence that the electroweak force and the strong force are parts of one grand unified force, and even that all forces including gravity are unified.

Another thing that lies behind the idea of a unique form for the laws of physics is that the structure of some parts of our present theory, the Standard Model, is found to be completely dictated by certain powerful principles—often, symmetry principles. A very good example is the part of the theory that describes how particles called "gluons" interact with each other by means of the strong force. The structure of these interactions is completely dictated by the symmetry called SU(3)×CP obeyed by the strong force. There is no choice left to be made once it is laid down that the strong force must respect this symmetry: the gluons must interact with each other in a unique way. The same thing is believed to be true about superstring theory, which many physicists think may be the ultimate unified theory of all physical phenomena. All the parts of superstring theory fit together so harmoniously that nothing can be changed without the theory ceasing to be mathematically self-consistent. No one yet knows what the profound underlying principles of superstring theory are. But whatever they are, they appear to completely dictate the mathematical structure of the theory.

As yet there is no direct experimental evidence that superstring theory is really the right theory of nature. One reason that many physicists nevertheless suspect that it is the right theory is that it is the only way that is presently known of reconciling the principles of Einstein's theory of gravity (i.e., the General Theory of Relativity) with the principles of quantum theory. If one attempts to apply quantum principles to Einstein's theory, the theory ceases to make mathematical sense: calculations that ought to give finite answers give infinite answers instead. In the technical jargon, Einstein's theory of gravity is "non-renormalizable." For decades, theorists tried to find a satisfactory quantum theory of gravity, but without success. It seemed an insuperable problem—until superstring theory came along. It may be that superstring theory is the only mathematically consistent way to combine quantum theory and Einsteinian gravity.

ANSWERS TO THE OBJECTIONS

1. The Requirements for Life

The first objection to the idea of anthropic coincidences was that we do not really know with certainty what the requirements for life are. And, indeed, this is quite true. We do have limited imaginations, and life conceivably could take forms utterly different from the life we know about. It could be based on different physics and arise in different ways.

What this means is that anthropic arguments will always have a question mark attached to them. They cannot achieve the rigor and certainty of mathematical demonstrations. However, this should not overly concern us. As I stated at the outset, and have had occasion to remind the reader since, this book is not about rigorous proofs. It is not a question of whose view, the theist's or the materialist's, can be rigorously proven from the scientific facts, but rather whose view is rendered more credible by the scientific facts. Materialists have long claimed, with great assurance, that the facts discovered by science render incredible the idea that the universe was designed with us in mind. If nothing else, the anthropic coincidences show this claim to be unjustified.

Moreover, while absolute certainty is not possible with regard to anthropic coincidences, a great deal of confidence is possible in some instances. For example, we cannot know exactly how long the universe had to last for life to have time to evolve, but we do know with virtual certainty that it had to last much longer than a "Planck time" (i.e., 10^{-43} second), since the very idea of time as we know it breaks down for anything shorter than that. Because the processes of organic life, and, even more, the gradual evolution of organic life, require enormously complex sequences of events, they must require a very large number of Planck times to unfold. This already implies that certain numbers such as the cosmological constant and the spatial flatness of the universe must be extremely small.

Other anthropic assumptions are not as certain as this, but are nonetheless very solid. For example, in chapter 15 we saw that if certain aspects of the laws of nature had been different only the element hydrogen would have existed, and it would have been unable to combine through nuclear fusion to make the nuclei of heavier elements (see examples 1 and 5). In that case, there would have been no chemistry to speak of, since the only chemical compound that hydrogen can form by itself is the hydrogen molecule, H_2, consisting of a pair of hydrogen atoms. Even though it has been suggested, I think few scientists would find it plausible that life could be made using only the chemistry of hydrogen. Furthermore, if hydrogen had been the only stable nucleus around, it is reasonably certain that nuclear processes could not have substituted for chemical ones as the basis of life. If all this is not bad enough for the prospects for life, without nuclear processes there would not have been stars. Now, it is true that even on Earth there are organisms that do not derive their energy from the Sun. For instance, there are bacteria, which live in volcanic vents deep under the ocean, that thrive off of geothermal energy. However, even geothermal energy comes ultimately from nuclear processes. It is radioactivity that keeps the earth's interior hot, and radioactivity comes from the decay of large nuclei. If only hydrogen nuclei had existed, this source of energy would have been just as impossible as energy from the Sun. So, while it is not totally inconceivable that life of some sort could evolve in an all-hydrogen universe, it seems extremely far-fetched.

To sum up, while we cannot know with absolute certainty what physical conditions would preclude the evolution of life, it is possible to make intelligent arguments about it and to reach some conclusions in which we can be reasonably confident. Since we are interested in the question of what is plausible or credible, rather than what is rigorously provable, such arguments are quite sufficient.

2. Conventional Scientific Explanations

The second objection was that many if not all of the facts that are cited as anthropic coincidences are likely to have conventional scientific explanations. I believe this to be true; however, I do not believe that it constitutes a real objection to the idea of anthropic coincidences. The essential point was well expressed by the astrophysicists B. J. Carr and M. J. Rees:

> One day, we may have a more physical explanation for some of the relationships . . . that now seem genuine coincidences. For example, the coincidence $\alpha_G \approx (m_e/m_w)^8$, which is essential for [nucleosynthesis], may eventually be subsumed as a consequence of some presently unformulated unified physical theory. However, even if all apparently anthropic coincidences could be explained in this way, it would still be remarkable that the relationships dictated by physical theory happened also to be those propitious for life.[14]

In other words, even if all the physical relationships needed for life to evolve were explained as arising from some fundamental physical theory, *there would still be a coincidence*. There would be the coincidence between what that physical theory required and what the evolution of life required. If life requires dozens of delicate relationships to be satisfied, and a certain physical theory also requires dozens of delicate relationships to be satisfied, *and they turn out to be the very same relationships*, that would be a fantastic coincidence. Or, rather, a series of fantastic coincidences.

If the objection we are considering is an attempt to argue that there is nothing coincidental in the anthropic coincidences, then it very obviously fails. However, there is another possible point to the objection. That point is that while there may really be anthropic coincidences, they do not imply that God or anyone else deliberately "arranged" things to allow life to evolve. The reason is that all of the things that might have been "arranged" were already inflexibly fixed by fundamental physical principles. Understanding it this way, the second objection really boils down to the same thing as the third objection.

3. Did God Have a Choice?

The third objection was, in essence, that God had no choice, and, therefore, that there was really nothing for God to do. He did not "decide" that the fine structure constant was to be small; rather the principles of grand unified theories required it to be small. He did not "decide" that space was to be flat; rather, the dynamics of cosmic inflation forced it to be flat. He did not "decide" that an energy level of carbon 12 was to have the right value to enhance the production of carbon in stars; rather, the locations of the energy levels of carbon are fixed by the structure of the fundamental theory of physics. This brings us to a critical question: How much choice was there in how the universe was put together?

To answer this, let us turn to the example that I have already used many times: the fine structure constant. In the Standard Model this is a free parameter; it could have been anything. But in a grand unified theory it is tied to various other parameters. Typically one finds that in simple grand unified theories the fine structure constant comes out to be quite small. If we find that the world is described by a grand unified theory, therefore, we will have at least a partial explanation of why the fine structure constant is a small number. However, that is not to say that "there was no choice" about what the fine structure constant could be.

In the first place, even though the value of the fine structure constant is tied to other parameters in a grand unified theory, many of those other parameters are free. For example, in grand unified theories there is a parameter called the "unification scale." This parameter has to do with how the unified force is broken apart into the separate electromagnetic, weak, and strong forces. If our universe is described by a grand unified theory, then there was a choice to be made about the numerical value of the unification scale. A different value of the unification scale would have meant a different value of the fine structure constant.

In the second place, there are an infinite number of different grand unified theories; so there was also a choice to be made among them. Many of them do not even have such a thing as an electromagnetic force. Others have no strong nuclear force. In fact, the kind and number of such "gauge forces" could have been practically anything, and so could their relative strengths, depending on which grand unified theory had been chosen.

Finally, there was the choice whether the structure of the universe would be that of a grand unified theory at all. There is no law of logic that says that it had to be. Mathematically speaking, as far as we know, a grand unified theory is no better than a non-unified theory such as the Standard Model.

Let us take another example: the spatial flatness of the universe. It is true that in a physical theory in which cosmological inflation takes place, the spatial flatness of the universe can be explained. However, the same points can be made as in our previous example. First, in a theory that gives cosmological inflation, the amount of inflation, and therefore the degree of spatial flatness that results, depends on the values of various parameters of the theory. Second, there are an infinite number of possible mathematical theories of cosmological inflation. And, third, there are also an infinite number of theories in which cosmological inflation does *not* occur. Thus, here also, there were many choices to be made by God.

But what are we to reply to those who suggest that the ultimate "theory of everything" will turn out to be so constrained by its fundamental principles that no room for choice exists within its framework? The reply is very simple: there was a choice of framework. It may be that the underlying principles of superstring theory, say, leave no further room for choice, but there was the choice of whether the universe would be based on the principles of superstring theory in the first place. It may be that the only framework that incorporates both the principles of quantum theory and the principles of Einstein's theory of gravity is superstring theory. If so, then a universe based on quantum principles and including Einsteinian gravity would have to be a superstring universe. But there was the choice of whether the universe would have Einsteinian gravity or not. And there was the choice whether the universe would be based on quantum principles or not.

When people say that the ultimate laws of physics will turn out to be "unique," what they really mean is that they will turn out to be the unique laws *that satisfy certain conditions*. But what conditions did the universe have to satisfy? And who decided that those conditions and not some other conditions had to be satisfied?

If we remove all arbitrary preconditions, and simply ask how many different mathematical structures there are that could serve as the laws of some hypothetical universe, the answer, quite simply, is infinite. The only thing we must clearly require of any hypothetical universe is that it be mathematically and logically self-consistent, and there is no limit to the number of different universes that would satisfy those minimal requirements. To make matters simple, let us stick to hypothetical universes that are based on classical rather than quantum principles. In the first place, such a universe could have any number of space dimensions. That already gives an infinite choice. Second, if the universe contains matter, that matter could be in the form of particles, or fields, or an infinite variety of other kinds of things (like strings or membranes). If the matter consists of particles, say, the universe could contain any number of particles — which gives another infinite choice. There could be one type of particle, or several types—even an infinite number of types—differing from each other in mass or in other properties. The forces experienced by particles could depend on their types, positions, or other properties in an infinite number of possible ways.

I have not even gotten into such questions as the geometry of space or space-time, and in particular whether it is Newtonian or Einsteinian. And I have limited myself to classical rather than quantum physics. If any theoretical physicist were paid a dollar for every possible universe he could think up, he could get rich very quickly. Would these universes be mathematically beautiful? For the most part maybe not. Would they be such that if they existed life would be able to evolve in them? Probably not, in the overwhelming majority of cases. But who says that those things should matter? The simple and absolutely undeniable fact is that the universe did not have to have the particular laws it does have by any sort of logical or mathematical necessity. In other words, God had a choice—in fact, an infinite number of choices.

17 Alternative Explanations of the Anthropic Coincidences

Once it is recognized that there are anthropic coincidences, there are three positions one may take with regard to them: (1) They *are* just coincidences, pure and simple. (2) The anthropic coincidences have a purely natural, scientific explanation. Or (3) the anthropic coincidences show that we were deliberately "built in from the beginning," presumably by God.

It is clear that possibility 1 can never be ruled out. Circumstantial evidence, even a mountain of it, can always be dismissed as merely coincidental by someone not disposed to believe it. However, a materialist who takes such a dismissive attitude is no longer arguing for his position from the evidence. He is admitting that the universe does indeed have the appearance of having been designed with us in mind. He simply chooses to dismiss that possibility from consideration. If the materialist is really to contend for his position, he must adopt position 2.

To understand how the anthropic coincidences might have a natural explanation, as position 2 asserts, let us think about the fact that Earth is a habitable planet. This itself is the result of many "coincidences." For example, Earth is not so close to the Sun that we burn up, but not so far away that we freeze solid. Earth is large enough to have sufficient gravity to retain an atmosphere, but not so large that it retains hydrogen gas, which would be the wrong kind of atmosphere, at least for us. One can proceed in this vein and find many things about Earth's characteristics that make it suitable for human life. However, there is a very simple and natural explanation for these "coincidences": there are a huge number of planets in the universe. Some are hot, some are cold. Some are large, and some are small. With about 100 billion stars in a galaxy, and about the same number of galaxies in the observable universe, and with a significant fraction of those stars presumably having planets, it is no surprise that the universe has some planets with the right conditions to support human life. In fact, we would expect there to be a large number of such planets.

Now, one might ask why it happens that "our" planet is one of the habitable ones. But that is not a good question, really. Obviously, we could hardly be living on an uninhabitable planet. To put it another way, some (maybe only one) of the habitable planets will evolve life, whereas, of course, none of the uninhabitable planets will. So whatever intelligent beings appear who are able to ask such questions will look around and necessarily see that they are living on a planet that has the right conditions for life. No coincidence.

The basic idea here is simple. If I try one key in a strange lock and succeed in opening it, that is an unlikely coincidence. But if I try a million keys, it is not surprising if one of them happens to work.

Now, some people have proposed that the same kind of idea can explain why our universe happens to be habitable. Suppose that there is a vast collection of universes, each having different laws of physics. Some might have three space dimensions, others less or more; some might have four basic forces of nature, others a different number; and so on. If there were a large enough number of types of universe in existence, then it would be no surprise if just by chance one of them had conditions propitious for life.

This idea goes by the unfortunate and misleading name of the Weak Anthropic Principle or WAP.[1] The name is misleading, because really there is no new "principle" involved here. What is being proposed is not some new principle, but simply the existence of many universes. The name is unfortunate because there are a number of different versions of "anthropic principles" floating around, some of which are deservedly regarded by most scientists who have heard of them as quite flaky. There are the Strong Anthropic Principle and the Final Anthropic Principle, for example. One wit has even joked about the "Completely Ridiculous Anthropic Principle." In order to avoid all the distracting irrelevancies associated with the name Weak Anthropic Principle, it would be nice to find a better name for the idea we are discussing, but that would probably only confuse matters more. So Weak Anthropic Principle or WAP is what we are stuck with.

It may seem as though we now have a standoff between the theist and the materialist as far as the significance of the anthropic coincidences is concerned. The believer can point to them as evidence that "we were built in," while the materialist can appeal to the Weak Anthropic Principle scenario as a way of explaining them without invoking God. I will return to the theism/materialism debate shortly. First, it is necessary to look a little more closely at the Weak Anthropic Principle.

THE WEAK ANTHROPIC PRINCIPLE: MANY DOMAINS

The Weak Anthropic Principle comes in two versions. In one version, all the different "universes" are in reality merely parts of one larger universe, and that one universe is the only one that exists. These different parts of the one universe can appear to have different physical laws, even though underlying all the apparently different laws is actually a single set of truly fundamental laws. This sounds like a bizarre notion, but in fact this kind of thing can happen in the kinds of theories that particle physicists and cosmologists have investigated in recent years.

One way this can happen is by means of the matter in different regions of the universe being in different "phases." We are all familiar with H_2O having the three phases of ice, water, and steam. Someone who had never seen ice melt or water boil might think that these were three different substances altogether, obeying different laws, but a chemical investigation would reveal that they are all made up of the same molecules obeying the same fundamental laws of physics.

In a similar way, the basic material constituents of our universe can be in different phases. For example, the world we live in contains protons and neutrons. However, if the temperature of matter were raised to about 3 trillion degrees centigrade, all the protons and neutrons would melt into a "quark-gluon plasma." An accelerator called RHIC has been built at Brookhaven National Laboratory on Long Island to create the conditions (however, only on a subatomic scale) where a quark-gluon plasma would be made. It is believed that at yet more fantastic temperatures, which cannot be created by human experimenters, but which existed in the first moments of the Big Bang, even electrons, quarks, photons, and the other particles of the Standard Model melt into some more exotic form of matter.

So far, I have talked about different phases existing at different temperatures. But they can also co-exist at the same temperature, as ice and water do at 0 degrees centigrade. In many theories of physics, the matter in different regions of the universe can be at the same temperature but in different phases. The physics in these different regions could look extremely different, including the kinds of particles and forces that existed in them. To a naive observer, it would look like these regions had different laws of physics altogether, just as to a naive observer ice, water, and steam might seem unrelated. A more profound investigation would show that matter in the different regions of the universe was really obeying the same set of fundamental laws. In the terminology of physics, it would be said that the regions had different "effective" laws, but the same underlying laws.

In a theory where this happens, these different regions of the universe are called "domains." A natural question is whether astronomers would see evidence of other domains if they existed. It depends. If these domains were within our horizon of about 15 billion light-years, they should have shown up in various ways. However, in theories where the universe has undergone "inflation" as described in chapter 7, these other domains would almost certainly be so far away from us as to be completely unobservable. In that case, the entire part of the universe that we can see, even with the most powerful telescopes, would lie within a single domain.

It should be mentioned that another way has been suggested whereby the universe could have many pieces with different "effective" laws. This way involves the so-called "many-worlds interpretation" of quantum theory. The many-worlds interpretation is something that I shall discuss in chapter 25. It should be emphasized here, however, that even if the many-worlds interpretation is correct it does not mean that different parts of the universe will have different effective laws; that is an added assumption that has to be put in.

In any event, whether the "domains" arise through different phases of matter or through quantum theory makes no difference for the issues that I am going to discuss in this chapter.

THE WEAK ANTHROPIC PRINCIPLE: MANY UNIVERSES

The second version of the Weak Anthropic Principle assumes that there really are many universes that have nothing whatever to do with each other. They are not parts or regions of a single universe. This may be hard to imagine. It certainly cannot be pictured. Perhaps a crude analogy will help. In the Many Domains idea, there is one universe, which can be thought of as being like a single book. The different domains are like different pages of that book. They are physically attached in some way. The pages have some relationship to each other. In the Many Universes idea, on the other hand, each universe is like a separate book, and the various books have nothing at all to do with each other.

The problem with this analogy, of course, is that even two separate books have some kind of relationship to each other. They may exist on the same shelf of a library, for example, or in the same city. In any event, one can talk about the distance that separates them in space, because both books exist within the same physical universe. However, in our analogy we are not supposed to think like that. The "books" are complete universes unto themselves. There is no overarching physical space or universe that contains them all.

The Many Universes idea is very different from the Many Domains idea in a number of important ways. In the Many Domains idea, all the domains, while having different "effective" physical laws, really, deep down, are governed by a single set of fundamental laws. That means that not every conceivable kind of domain exists, but only the kinds of domains that are allowed by the fundamental laws of the one and only universe of which they are a part. For example, in the multi-domain theories that physicists have studied, the law of conservation of energy is generally assumed to be valid in every domain, and the number of space dimensions is usually assumed to be the same in every domain.

In short, the amount of variety that exists among domains in a Many Domains scenario is restricted by the nature of the fundamental laws that govern the one and only universe. On the other hand, if there are really many universes which have nothing whatever to do with each other, each with its own fundamental laws, then there can be no law or principle which governs how much variety there is among them, or how many there are. There might be two universes. There might be ten. There might be an infinite number. Some thinkers have gone so far as to suggest that every logically possible universe exists.

Another difference between the two scenarios is that in the Many Domains scenario, there is at least some hope that someday someone might be able to infer the existence of the other domains indirectly. This could be done if the fundamental laws of nature are discovered, and experimentally proven to be correct, and it is found that they mathematically imply the existence of other domains. (This is a long shot, but not out of the question.)

In the case of Many Universes, on the other hand, we could never know by any observations or inferences drawn from observations that the other universes existed. The only way one could ever know that they existed would be by a divine revelation.

Not surprisingly, physicists who advocate the Weak Anthropic Principle as an explanation of the anthropic coincidences almost all think in terms of Many Domains scenarios rather than Many Universes.

The Weakness of the Weak Anthropic Principle

I referred earlier to an apparent standoff between the theist and the materialist when it comes to the anthropic coincidences. The first thing that should be said about this standoff is that, even if it cannot be resolved, it would give the lie to one of the materialist's main claims, which is that the scientific evidence points to the insignificance of man in the cosmic scheme. If there really is a standoff, then the evidence is not pointing one way or the other, and "scientific materialists" cannot claim to be any more scientific, at least on this issue, than theists.

But is there really a standoff? More precisely, does the possibility of explaining anthropic coincidences by Many Domains or Many Universes scenarios really nullify their value as an argument in favor of theism? I am convinced that the answer is no.

I should make clear, to begin with, that I regard the Many Domains idea as a perfectly reasonable one for explaining anthropic coincidences, and I think it not unlikely that some of the coincidences will eventually be explained in this way. In fact, I am one of the co-authors of a paper published in *Physical Review Letters*, entitled "Anthropic considerations in multiple domain theories and the scale of electroweak symmetry breaking."[2] Nevertheless, I do not believe that Many Domains scenarios could ever completely draw the teeth of the anthropic coincidences as an argument in favor of theism.

The basic point of the anthropic coincidences, for the theist, is that they highlight the fact that the universe might have been a different sort of place, and that it had to be a very special sort of place if it were to be able to give rise to life. The Many Domains scenarios do not refute this contention. The reason they do not is that a universe which gives rise to a multiplicity of domains is itself a very special sort of place.

Suppose that I want to explain, using a Many Domains scenario, the fact that the parameter v has the value that it does, as seems to be required for life. (See example 5 in chapter 15.) To do this I have to assume that the parameter v varies from one domain to another. Not only does it have to vary, but it has to sample an almost continuous range of values. This is a tall order. It is not so easy to cook up theories of that kind. In virtually all the theories that anyone has studied, even ones having a domain structure, the parameter v does not vary among the domains in this almost continuous way.

Moreover, explaining v is not enough. There are many other anthropic coincidences: the value of the cosmological constant, the number of space dimensions, and so on. If the Many Domains idea is to explain all of them, then not only v, but the value of the cosmological constant, the number of space dimensions, and many other features of the effective laws of nature must vary among the different domains of the universe. This is an even taller order. A universe like that would be, to put it mildly, very special indeed.

And what if the laws of nature had *not* been such as to give rise to this very special sort of universe? Then, presumably, the precise conditions needed for life would not have been hit upon, and life would not have arisen. In other words, having laws that lead to the existence of domains of a sufficiently rich variety to make life inevitable would *itself* qualify as an anthropic coincidence.

There seems to be no escape. Every way of explaining anthropic coincidences scientifically involves assuming that the universe has some sort of very special characteristics that can be thought of as constituting in themselves another set of anthropic coincidences.

The Problem with Too Many Universes

Now, let us turn to the Many Universes version of the Weak Anthropic Principle. This idea has almost no following among scientists, but it has an advantage for the materialist. In the Many Domains idea, it must be assumed that the laws of physics are engineered in just the right way to produce a great number of domains having a sufficiently rich variety of conditions that life is bound to arise in some of them. In the Many Universes idea, however, such clever engineering is not required. The many universes with all their rich variety are just *supposed* to exist, by fiat, as it were. (Whose fiat? We are not supposed to ask.) There is no overarching law that explains scientifically why some kinds of universes exist and some kinds do not. If there were such an overarching law, then the different universes would be governed by this common law, and would in essence be simply domains of a single universe.

So, in the Many Universes idea, the universes that exist just happen to exist, and the universes that do not exist just happen not to exist. This raises inevitable questions. Why do some kinds of universes exist and others not? If, say, three-dimensional universes exist, but no seven-dimensional universes exist, why is that? If universes with gravity exist and universes with no gravity do not exist, why is that? It sounds very much like someone is making choices here. Far from escaping from the idea of a cosmic designer, this scenario seems to make the case for one much stronger.

There is, however, a last way out for the materialist, though few of them are bold enough to embrace it. The way out is to grasp the nettle and assume that all conceivable universes exist. More precisely, it is to assume that any universe that is logically and mathematically self-consistent actually exists. This certainly would explain why habitable universes exist: habitable universes exist because they are logically possible.

There is a beautiful simplicity to this idea, which is called "modal realism" by philosophers. It completely avoids the necessity of anyone making any choices. No need, therefore, for a designer. This is the opposite extreme from the idea that the laws of physics are unique, discussed in chapter 16. There the idea was that no designer is needed because there is only one possible kind of universe and therefore no choices to be made. Here it is suggested that there is an infinity of possible universes, and hence an infinite number to choose from, but that no one has actually chosen among them, so that they all equally exist.

From a philosophical point of view, this idea is very interesting. One thing that it does is to abolish the distinction between the possible and the actual. Anything that is possible is actual. It also gives an answer to the philosophical question of what it means to "exist": "to exist" means the same thing as "to be self-consistent."

While it has the virtue of a kind of simplicity, most people would regard this idea as somewhat crazy. And when one thinks about its full implications, crazy does not seem too strong a word for it. If all possible universes exist, then there is a universe where *The Wizard of Oz* is a true story, and another where Kermit the Frog is a real person. It is not surprising that very few people have ever adopted modal realism. In fact, the late American philosopher David Kellogg Lewis may be the only one.[3] That the idea seems crazy, however, should perhaps not be held too much against it. There is a famous true story about the great physicist Niels Bohr, who objected to a wild new theoretical idea not because it was crazy, but because it was "not crazy enough." His point was that the world is a very strange place, and for that reason may not be explicable except by ideas that depart very far from our ordinary notions of what is "common sense." That was certainly true of quantum theory, the great theory which Bohr himself helped to found.

But possible craziness aside, the idea that all possible universes exist has a fatal flaw. It makes it impossible to understand the orderliness of nature.

If all possible universes exist, then there will certainly be universes that, like ours, obey with perfect precision a set of beautiful and orderly mathematical rules. However, universes would also exist that obey a set of rules most of the time, yet suffer occasional or frequent exceptions to them. There would also be universes that obey different rules at different times, and yet others that are so haphazard that hardly any general rules could be found to apply to them. Indeed, one would expect universes which are as consistently lawful as ours to be extremely exceptional. Why, then, do we live in such a universe? A possible answer, but one which falls short, is that a universe must be lawful to some degree if it is to be able to evolve such complex structures as ourselves. In other words, any universe which harbors life and intelligence is bound to be exceptional. However, this cannot explain why our universe has the perfection of lawfulness that ours appears to have.[4]

This objection to the idea that all possible universes exist is the same as the objection made in chapter 13 to the idea that the laws of nature are a result of chance. This is not surprising, since if all logically possible universes exist the lawfulness of our universe *is* a result of chance.

We come down to the following basic conclusion: Merely postulating the existence of an infinite number of domains to our universe or altogether separate universes cannot get around the fact that our universe is a special kind of place—indeed, doubly special. Among all possible universes, ones that allow for life and intelligence to evolve are exceptions, and those that also have the perfection of order which ours exhibits are exceptions among the exceptions. It is difficult to escape the impression, even if it may be hard to prove with philosophical rigor, that a choice was made. Or, rather, many choices. How did those choices get made, or why? Materialism can give no answer.

In any event, however inadequate, some version of the Weak Anthropic Principle seems to be the only way to attempt to explain the anthropic coincidences in a naturalistic way. It is a very curious circumstance that materialists, in an effort to avoid what Laplace called the unnecessary hypothesis of God, are frequently driven to hypothesize the existence of an infinity of unobservable entities. We saw this before, in chapter 7, with the speculation that an infinite time preceded the Big Bang. We saw it again in the first part of chapter 11, with the idea that the first living thing might have arisen by chance if the universe is infinitely large and has an infinite number of planets. We see it now, in the idea of a large and possibly infinite number of domains or universes. We shall encounter it once more, in chapter 25, with the so-called many-worlds interpretation of quantum theory. It seems that to abolish one unobservable God, it takes an infinite number of unobservable substitutes.

18 Why Is the Universe So Big?

As we have just seen, a strong case can be made from the mathematical structure of the laws of physics that we were "built in from the beginning." But if on some clear night we go outside and gaze up at the heavens, it is natural for us to be shocked at their immensity and overcome by a sense of our own insignificance. How vast the universe is, and how very small we are! How can we be so arrogant as to think that all of this was made for our sakes? Is that not as ridiculous as thinking that the entire Pacific Ocean exists for the sake of one microbe floating in its depths?

Surely, there is at least a paradox here. It is a paradox that was not lost upon the Psalmist, who cried out,

> When I consider the heavens, the work of thy fingers, the moon and the stars, which thou hast ordained;
>
> What is man, that thou art mindful of him? and the son of man, that thou visitest him? (Ps. 8:3–4)

Doubtless it was thoughts such as these which impelled Blaise Pascal to confess in his *Pensées*, "The eternal silence of the infinite spaces frightens me."[1]

But is this fear and sense of insignificance, however natural it may be, completely rational? Is it not based on an unthinking tendency, which is really absurd, to equate size and significance? Who would deny the foolishness of saying that because one thing is a thousand times bigger than another it is also a thousand times more important? No, size is not a measure of greatness or importance. What sets human beings above all the stars and galaxies revealed by our telescopes is our power of reason. The universe, for all its immensity, knows nothing and understands nothing, while our thoughts and understanding can reach from one end of the cosmos to the other. Because of the power of our minds we can each, with Hamlet, say, "I could be bounded in a nutshell and count myself king of infinite space."

Even Bertrand Russell, who, as we have seen, called the human race "a curious accident in a backwater," was convinced that human beings dwarf the whole rest of the universe in value. He wrote,

> I have long ardently desired to find some justification for the emotions inspired by certain things that seemed to stand outside human life and to deserve feelings of awe. I am thinking . . . in part of the vastness of the scientific universe, both in space and time, as compared to the life of mankind. . . . And yet I am unable to believe that, in the world as known, there is anything that I can value outside human beings. . . . Not the starry heavens, but their effects on human percipients, have excellence; to admire the universe for its size is slavish and absurd."[2]

Einstein expressed the same view with even greater vehemence. He has been quoted as saying, "If there were not this inner illumination [i.e., the human mind], the universe would be merely a rubbish heap."[3]

Even though I believe Russell and Einstein went too far in seeming to deny the admirableness of the universe apart from human beings, they are surely right that it is foolish to deny human importance merely on account of size. Nevertheless, a question remains. Why *is* the universe so large? If it was really made for our sake, were all those "infinite spaces" necessary? The religious person would reply, of course, that the universe was not made only for our sake. Neither Jewish nor Christian tradition makes that claim. Indeed, in the account of Creation given in the Book of Genesis, after each part of the material world is created by God, the words are repeated, "And God saw that it was good." This is said before the creation of man is recounted, and so the implication is plain that God sees these things as good, not only in relation to us, but in themselves. Another theme running through the Bible is that the greatness and majesty which we see in the physical universe show forth God's own greatness and majesty: "The heavens declare the glory of God," says Psalm 19. But if they declare God's glory rather than our own, there is no reason to expect them to have a merely human scale. In the fifteenth century, indeed, the theologian Nicolas of Cusa suggested that the universe might be infinite in size precisely because he thought that such a universe would better reflect the infinity of its creator.[4]

There is, however, another way of answering the question about the vastness of the universe, both in age and in size. It comes from physics rather than religious or philosophical reflection. As we shall see, the very vastness of the universe seems to be necessary if life is to arise in it, and so may actually underline rather than contradict the importance of human life in the scheme of nature.

How Old Must a Universe Be?

The universe is indeed very old. The current best estimate of its age is about 15 billion years. This is a long time in human terms, but it is not long compared to what is required for life to arise through evolution. In the first place, life as we know it requires complex chemistry, and most of the elements that are used in the chemistry of life were forged in the deep interiors of stars. In order to become available for life, these chemicals had to be released by the stars in which they were formed. This happened at the end of those stars' lives, when they exploded as supernovas. Before biological evolution could even commence, then, the whole lifetime of a star, which is typically billions of years, had to elapse. The Sun, a later generation star, began its life about 10 billion years after the Big Bang.

The process of biological evolution, too, is very time-consuming. Evolution is not understood well enough to be able to say how long it must take in general. But we do know that on Earth it took billions of years to produce multi-celled creatures, and several hundreds of millions of years more to produce animals with complex brains.

Now, these are facts about our own universe. However, one can argue that in any universe which could evolve sentient life through natural processes the universe itself would have to last much longer than the lifetimes of individual creatures. In the first place, a sentient organism is necessarily a complex thing. Its brain must be composed of a large number of elementary parts. This is certainly so of the human brain, which contains about 100 billion neurons, each of which is itself an intricate structure. The brain of a sentient creature might be very different from a human brain, of course, but it can hardly be doubted that, in general, it would have to possess a huge number of parts. Consequently, if such a structure is to be built up by a gradual process of evolution, a great many steps are required. This, in turn, implies that many generations must unfold. One generation, therefore, must be a small part of the total age of the universe.

How Big Must a Universe Be?

In our universe, because of its expansion, the size of the universe is directly related to its age. As a consequence of the fact that the universe has been around for 15 billion years or so, it has grown to a size of at least 15 billion light-years. That is a staggering 10^{26} times bigger than a human being. However, if the universe were any smaller than that it would not have lasted long enough for life to have evolved. This is an absolutely crucial fact.

For example, suppose that we wanted the universe never to get larger than a million miles across, a more comfortable and less intimidating size. What this would imply—assuming Einstein's equations held good—is that the universe would not have lasted more than a few seconds. The universe would have expanded to a million miles across in a matter of seconds after the Big Bang, and then collapsed upon itself. Even a universe whose greatest size was that of the whole Milky Way would be very short lived: it would last only a few hundred thousand years, certainly not enough time for life to evolve.

There is another reason why the universe might have to be very large in order to have life. If the origin of life involves random chemical processes, then a huge number of chances have to be allowed for the right combinations to occur. It is beyond anyone's ability at the present time to estimate these numbers, even roughly, but they certainly have to be tremendously large. This suggests that if life arose in our universe through chance processes, it probably required a huge number of planets for there to be even one where life was able to get a start.

Of course, these arguments assume certain things about the nature of the universe. The first argument assumed that the size of the universe is determined by Einsteinian gravity, while the second argument assumed that life began through chance chemical interactions. Other possibilities than these can be imagined. But what these arguments show is that at least under some assumptions the formation of life can require a universe that is vastly larger than the size of the living things it contains.

ARE WE REALLY SO SMALL?

Whether we regard something as large or small clearly depends on what we compare it to. If human beings are compared to the entire universe, then we come out looking pathetically small. But in comparison to atoms, we look quite huge. In fact, rather than being small in any absolute sense, human beings stand near the midpoint between the very large and the very small in nature. To see this more clearly, let us look at some numbers.

For the very small, let us take an atom. An atom is not the smallest thing in nature by any means, but it is a very basic building block. For the very large, let us take the observable part of the universe. The size of an atom is about an Angstrom unit, which is one hundred-millionth of a centimeter, or in scientific notation 10^{-8} of a centimeter. The observable universe is about 15 billion light-years across, which comes out to roughly 10^{28} centimeters. If we ask what distance is the "geometric mean" between these two extremes—that is, the length which is as many times larger than an atom as the universe is larger than it—one comes out with a distance of about 10^{10} centimeters, which is the size of a large planet. In other words, the universe is to a planet as a planet is to an atom. Now, if we ask what distance is the geometric mean between the size of a planet

and the size of an atom, we find that it is about 10 centimeters, a very human scale indeed—about the width of a human hand.

These are no accidents, as it turns out. One can estimate how large an animal can be, how large a planet is, and how large the universe is, using rather basic physical considerations.

How Basic Physics Sets the Relative Sizes of Things[5]

Consider first planets. There is a largest size that a planet can be if it is not to get crushed by its own gravity and become a star. And there is the smallest a planet can be if it is not to be really an asteroid, an irregularly shaped object with no atmosphere. What one finds is that a typical planet has a size that is about the square-root of the ratio α/α_G times the size of an atom. In this expression, α is the "fine structure constant" that measures the strength of the electromagnetic force, and α_G is the "gravitational fine structure constant" that measures the strength of the gravitational force.[6] When one plugs in the numbers this rough theoretical estimate turns out to be very close to the actual typical size of a planet.

The size of the universe is a bit trickier. First of all, we cannot be sure how large the entire universe is, since we cannot see all of it. Moreover, both the universe and the observable part of it are expanding. What I mean by "the size of the universe" here is the diameter of that part of it which is observable at the time when life first appears. Now, for reasons explained earlier, living creatures cannot appear until at least some stars have had time to explode and release the heavier elements needed for biochemistry. One can estimate when this happens, since the life cycle of stars is rather well understood. This calculation then allows one to figure out a minimum size of the universe. The minimum size of the universe turns out to be about the ratio α/α_G times the size of an atom. When the numbers are plugged in, this theoretical estimate is found to be quite close to the actual size of the observable universe at the present time.

The upshot is that it is possible to understand in terms of some basic physics why planets are as big as they are and why the universe is as big as it is. These sizes are not arbitrary. And what the basic physics tells us is that the size of a planet must be, roughly speaking, the geometric mean between the size of an atom and the size of the universe. In other words, this relationship is not an accident, but is fundamental to the way our universe is constructed.

The same thing is true about the size of living things. One can use basic physics to set a limit on how big a living thing can be, at least if it lives on land. The idea is that a land animal cannot get to be too large without its weight becoming a problem for it. An animal's weight depends, of course, on where it is living: it would weigh much less on an asteroid or on the Moon than on Earth. But in our calculation we will assume that animals live on planets, and since we have already estimated how big planets are, we can figure out how strong their surface gravity is. However, we still need to know when an animal's weight becomes a problem for it.

Now, obviously, some things on the surface of planets—such as skyscrapers and mountains—can be very large and still support their own weight. However, skyscrapers and mountains do not move around much. Because animals do move around, they sometimes fall down, and it seems reasonable to assume that animals that would break apart on falling down would not be very competitive in the struggle for survival. Arguing in this fashion, one can set a very rough upper limit to land animal sizes. The answer does indeed come out to be about the geometric mean between the size of an atom and the size of a planet. (One can only get very rough estimates this way, because a lot depends on the kinds of chemical bonds that hold the animal together.)

One can play the same kind of games with lengths of time. An atomic time scale can be taken to be the time it takes an electron to orbit the nucleus. This is about 3×10^{-17} of a second. The age of the universe is about 5×10^{17} seconds. The geometric mean between these extremes is 4 seconds—a very human time scale indeed. A heartbeat is about 1 second, and it also takes about 1 second for the brain to initiate a voluntary bodily action.

Again, this is not too surprising. The human body is a very complicated physical system. A basic biological process, like a heartbeat or an action of the brain, involves a vast number of chemical processes to happen in sequence. "Human time scales," therefore, must be very long compared to the basic atomic time scale. On the other hand, as I discussed earlier, the age of the universe has to be much longer than the lifetime of a living thing, and therefore much longer still than the time required for a basic biological process.

The Golden Mean

Alexander Pope's *Essay on Man* contains the famous lines,

> Know, then, thyself. Presume not God to scan;
> The proper study of mankind is man.
> Placed on this isthmus of a middle state,
> A being darkly wise and rudely great.

The poem goes on to contrast the greatness of man with his limitations, the reasons both for human pride and humility. What science shows us is that even on the physical level human beings are indeed placed on an "isthmus of a middle state." We are perched at the halfway point between the very largest and the very smallest things in nature, and between the longest and the shortest times.

To the scientist, this is not surprising. The "middle state" is where the most interesting phenomena often happen in physical systems. When looked at on the very longest scales of time and distance, things are often rather dull, because all the interesting details have been glossed over. Looked at on the smallest scales things look simple because there is not enough scope for complexity to arise. An atom is simple to describe, but so, in a way, is the universe as a whole — at least as the cosmologist sees it. The cosmologist does not care about what individual stars or even galaxies or clusters of galaxies are doing. He only cares about the "big picture"—the biggest of big pictures.

If you look at the whole forest, or just at a single leaf, you will miss the most interesting things that are going on, which happen in between. "In between" is where human life is lived out. As Copernicus showed, we are not at the geometric center of the universe. But from the point of view of scales of time and distance, we are in a very real sense in the middle.

PART V

What Is Man?

19 The Issue

In the last several chapters we have been concerned with the question of the place of man in nature; but this raises a more fundamental question: What is the nature of man? To this the materialist and the religious believer give radically contradictory answers. The materialist says that human beings, like everything else, are merely complex material systems, and are therefore no more, in essence, than extremely clever animals. The Jewish or Christian believer claims, on the contrary, that there is something in our nature—the spirit or soul—which sets us apart in a fundamental way from all that is purely physical. The question, then, is not merely what the place of man is in the physical world, but whether we are indeed just parts of the physical world.

That there is something different about man is obvious. Everyone, whether materialist or religious believer, can see that there is an enormous gulf between the capacities of human beings and animals. Animals do not spend time wondering about their place in nature or their fate. They do not write works of philosophy. They do not study chemistry or do research in astrophysics, compose violin concertos or paint with watercolors. They do not build ships or airplanes, tell time with calendars or clocks, cultivate gardens or erect magnificent buildings, adorn their bodies with jewelry and clothing or their dwellings with works of art. They do not contemplate what is ironic or tragic or humorous in life. They do not devote themselves to charitable works, or pray to God, or perform religious devotions. They do not deliberate in congresses or courts of law. They do not engage in agriculture, manufacturing, finance, or commerce. The lives of animals are simple and unvarying, while the list of distinctively human activities is virtually endless.

The difference between man and animal is also moral. Animals do not contemplate the morality of their acts or suffer pangs of conscience. Moral categories simply do not apply to them. The animal world is the scene of constant slaughter; yet we do not pass judgment on the jackal or the hawk. Nor do we account the lives of animals as having the same moral status as our own, or even as capable of being weighed in the same scales with ours. Though we may sympathize with the sufferings of animals, we do not regard a man who kills an animal, however wantonly or cruelly, as having committed the crime of murder.

All perceive that human beings and animals appear to exist on different planes.

The Religious View

That there is an enormous gap between man and animal is undeniable; but what is the basis of it? In the eyes of Jews and Christians, the basis of it is spiritual. In the biblical account of creation, only of man is it said that God breathed into him the breath of life and that "he became a living soul" (Gen. 2:7). This has always been understood to symbolize the fact that only upon human beings, among all the living things of Earth, did God confer a spiritual nature similar in some way to his own. Only man is said to be made "in the image of God" (Gen. 1:26–27).

But in what way are we made in the image of God? And in what way do we have a spiritual nature? According to the traditional Jewish and Christian doctrine, we are made in the image of God primarily in the sense that, like God, we have rationality and freedom. For example, we find in the second century St. Irenaeus of Lyons writing, "Man is rational and therefore like God; he is created with free will and is master over his acts."[1] When religious believers say that we have a "spiritual soul," therefore, they are not referring to something occult or magical; they are referring to the faculties of intellect and free will that are familiar to and constantly employed by all human beings. In the words of Calvin, "God [exalted] man above all the other animals to separate him from the common number, because he has attained to no vulgar life, but a life connected with the light of intelligence and reason, — [this] at the same time shows how he was made in the image of God."[2] The *Catechism of the Catholic Church* expresses the same ancient doctrine: "By virtue of his soul and spiritual powers of intellect and will, man is endowed with freedom, an 'outstanding manifestation of the divine image.'"[3]

According to theology, these "spiritual powers of intellect and will" are "transcendental," in the sense that they allow human beings to transcend or "go beyond" themselves and their own senses, appetites, and desires, so as to be able to perceive what is objectively true, beautiful, and good. The religious believer claims that these powers not only exceed the endowments of the lower animals, but necessarily exceed the capacities of anything that is purely physical. In the words of the same *Catechism*, "[In] his openness to truth and beauty, his sense of moral goodness, his freedom and the voice of his conscience, . . . [man] discerns signs of his spiritual soul. The soul, the 'seed of eternity we bear in ourselves, irreducible to the merely material,' can have its origin only in God."[4]

THE MATERIALIST VIEW

Of course, in the view of the materialist there is no such thing as a "soul," "spirit," or "seed of eternity"; there is only matter. According to the materialist, a human being, no less than a frog or a rosebush, is a complex aggregate of subatomic particles arranged in a hierarchy of structures, from atom, to molecule, to cell, to organ. That is the whole of human reality. It may turn out that the subatomic particles are composed of some yet more fundamental constituents, but that would not change the basic picture: the human animal, like all animals, is conceived of by the materialist as being merely a complex physical system. To know all the parts of that physical system and how they are moving through space and time is to know, in the final analysis, everything that there is to know about the human being.

This reductionism may affront our feelings or our intuitive beliefs, but the materialist sees the entire course of scientific history as leading inevitably toward this conclusion. In particular, the materialist would point to five developments.

Biology Has Been Shown to Be Reducible to Chemistry and Physics

There was a time when many scientists felt that there was a divide separating the world of inanimate things from the world of living organisms. It was thought that in order to explain life one would need laws and principles of a fundamentally new kind. This point of view was called "vitalism." Some spoke of a "vital force," or "élan vital." It was not only religious people, by any means, who entertained such notions. For example, George Bernard Shaw, who was not religious in any traditional sense, wrote about "The Life Force."

All this is now seen to be wrong. At least, it is wrong at the microscopic level. There are still a few people who maintain that at a higher level of description, above atoms and molecules, there may come into play some kind of "holistic" laws or "psycho-physical" laws or something else of that kind. But such ideas are generally dismissed by researchers in the physical sciences. Ever since 1828, when Friedrich Wöhler artificially synthesized urea, a major component of urine, it has become increasingly clear that the chemistry that goes on in the bodies of living things and the chemistry that goes on in inanimate objects involve the same basic elements reacting with each other according to the same basic laws. A great deal is now understood about the processes of life — metabolism, respiration, the immune system, the nervous system, reproduction, and so on—and there is no hint of anything happening that is not derivable from the same principles of physics that apply everywhere else.

Man Appears Physically to Be an Animal like Other Animals

This was generally recognized long before the theory of evolution came along, of course, but that theory has rubbed our noses in the fact. It is now generally accepted that *Homo sapiens* evolved from now-extinct ape-like creatures, the oldest remains of which have been discovered in East Africa. A commonly given sequence is *Australopithecus afarensis, Homo habilis, Homo erectus,* and *Homo sapiens.* There are respected biologists who even argue that the gorilla and chimpanzee should be classified in the genus *Homo.*[5] Certainly, by some genetic measures, these animals are physically very close to humans. For example, the DNA sequences of chimps and humans overlap by about 98 percent.

The gap between animals and humans has also come to appear somewhat narrower because of research in animal intelligence. It has been found that some animals do use and even make tools. For example, chimpanzees will strip the branches and leaves off of sticks so that they can be inserted into holes to draw out ants or termites for food. And chimpanzees and gorillas have been taught to have some rudimentary ability to use sign language.

There Is a Correspondence between Physical Events in the Brain and Mental Processes

The progress in neuroscience and the study of the brain have shown just how close this correspondence is. The brain and the functions of its different areas have been mapped out in some detail using a variety of techniques. These include studying people whose brains have had parts accidentally damaged or surgically removed, observing the effects of the electrical stimulation of specific areas of the brain, and measuring patterns of neural activity or other changes in the brain while subjects are performing specific mental tasks. Many discoveries have dramatically illustrated the physical basis of various mental phenomena. For example, a gifted musician who suffered a vascular lesion to the right temporal lobe of his brain retained all of his technical musical skill, but apparently lost any aesthetic feelings associated with his performances.[6] And recent experiments have found that electrical stimulation of a certain small region of the brain produces reactions in experimental subjects ranging from mild amusement to hilarity, depending on the strength of the electrical current employed. ("At low currents only a smile was present, while at higher currents a robust contagious laughter was induced."[7])

Computers Have Shown that Machines Can Perform "Mental" Tasks

Computers can receive data about the world around them from sensory devices, encode this information, analyze it, store it in memory, retrieve it, use it to make decisions, and put those decisions into effect by means of other devices controlled by them. Machines can, in a sense, thus be said to perceive, to choose, and to do. They can perform very well at certain tasks that, for humans anyway, require intelligence, such as playing chess well.

It has become a widespread opinion among neuroscientists and philosophers of mind—and seems to be a dogma in the field of artificial intelligence (AI)—that the human mind is nothing more than a functioning computer. It is true that the human brain is vastly more complicated than any computer built by humans, and uses neurons rather than silicon chips, but it is claimed by some that it can do nothing that an electronic digital computer could not—in principle—do. The most enthusiastic proponents of this view believe that the day is not far off when human-built machines will actually think and become conscious. (Whether "androids" will ever act like people, the plausibility of this idea in the mind of the general public has been greatly increased by movies and television shows in which people play androids.)

What Happens in the Physical World Is Rigidly Determined by Physics

All of the physical world is subject to precise mathematical laws, which, in their classical form at least, are "deterministic." They are deterministic in the sense that if the state of the physical world were completely specified at one time then the state of the world at any future time would be completely determined and could therefore—in principle—be calculated. One of the first to emphasize this point was the great French physicist and mathematician Laplace. In 1819, in his *Philosophical Essay on Probability*, he wrote, "[For] an intelligence which could know all the forces by which nature is animated, and the states at some instant of all the objects that compose it, nothing would be uncertain; and the future, as well as the past, would be present to its eyes."[8] Most people have recognized that this, if true, would be fatal to the doctrine that human beings have free will. The point is that, even if our minds were able to make free choices, these choices could have no effect on the actions of our bodies, because our bodies are made of matter and would have to behave in the way determined by the prior state of the material world.[9]

Some have claimed that the new science of "chaos theory" breaks the grip of determinism. This is false. What chaos theory does is show that a very small uncertainty in the "initial conditions" of a physical system can lead to very large uncertainties in the system's subsequent behavior. Because of this sensitivity, and because we can never measure the state of the world or even of a small part of it exactly, the future is highly unpredictable in practice. But being unpredictable in practice is not the same as being undetermined in principle. In classical (i.e., non-quantum) physics, chaotic systems are every bit as deterministic as any other systems. Chaos theory alone, therefore, is not enough to rescue free will.

For the materialist, the trend is clear: every development in science seems to narrow the gaps between human, animal, and machine. Every advance seems an advance toward a mechanistic view in which there is no room for any element in man that is "irreducible to matter"—any "ghost in the machine" as it has come to be derisively called.

CLEARING UP SOME CONFUSIONS

Before getting down to the real issues that divide materialist and believer, it is necessary to dispose of some false ones. There is no question that all five of the scientific developments I have just discussed are consistent with the expectations of materialists. What is often not realized, however, is that most of them are also entirely consistent with the expectations of religious believers. Most of them, therefore, have little relevance to the debate between religion and scientific materialism.

For example, it has never been a part of Jewish or Christian doctrine that the human body differed in any essential way materially from animal bodies or inanimate matter. In fact, the symbolic language of the book of Genesis teaches quite the opposite when it says that man was formed from the "dust of the ground" (Gen. 2:7). Some biblical literalists have interpreted this to mean that we were fashioned directly from dust, rather than arising through a gradual process of evolution. But in either case, whether directly or gradually, the Bible certainly claims that it is from this dust that we ultimately came. This point is made repeatedly throughout the Bible, as in Psalm 103: "Like as a father pitieth his children, so the Lord pitieth them that fear him. For he knoweth our frame; he remembereth that we are dust." (Ps. 103:13–14) This is something that Christians from time immemorial have been called on to remember as well. On every Ash Wednesday, as ashes are placed on their foreheads as a call to repentance and a reminder of death, Christians have been admonished, "Remember, man, that thou art dust, and unto dust thou shalt return."

Nor is it in any way contradictory to the traditional views of Jews and Christians that from the physical standpoint we are animals, and that our bodies are like the bodies of other animals. This point could not be made with more brutal frankness than it was in the Book of Ecclesiastes:

I said in mine heart concerning the estate of the sons of men, that God might manifest them, and that they might see that they themselves are beasts. For that which befalleth the sons of men befalleth beasts; even one thing befalleth them: as one dieth, so dieth the other; yea, they all have one breath; so that a man hath no preeminence above a beast: for all is vanity. All go unto one place; all are of the dust, and all turn to dust again. (Eccles. 3:18–20)

Jews and Christians have never been embarrassed by the fact that human beings are animals. The mediaeval Scholastic theologians accepted the Aristotelian definition of man as "a rational animal."

Consider now the third discovery, that there is a close correspondence between physical events in the brain and mental phenomena. This does not contradict the religious view, because the religious view does not deny the link between the physical and mental. What the religious view does deny is that *certain* of the capacities of human beings—specifically their intellect and free will—can be *entirely* reduced to the material. The word *entirely* is important here. Religious people do not deny that the brain is *necessary* for much and perhaps for *all* mental functioning, they only deny that it is *sufficient* for *some* mental functioning. This is frequently overlooked (or forgotten) by even quite careful writers on the subject. For example, the mathematician and physicist Roger Penrose, in his book *Shadows of the Mind*, asks, "If mentality is something separate from physicality, then why do our mental selves seem to need our physical brains at all?"[10] But the religious view, which Penrose is questioning here, does not claim such a separateness. The *Catechism of the Catholic Church*, for example, rejects it explicitly: "The unity of soul and body is so profound that one has to consider the soul to be the 'form' of the body. . . . Spirit and matter, in man, are not two natures united, but rather their union forms a single nature."[11]

It was obvious long before the investigations of modern neuroscience that physical events in the brain affect thinking. A sharp blow to the head can cause unconsciousness. Too much alcohol causes drunkenness. Fatigue makes it hard to concentrate. It is not surprising that electrical stimulation of a part of the brain can cause hilarity, when tickling someone in the ribs can have the same effect. That there is a connection between physical and mental events, then, has always been clear, although it must be admitted that modern researches have made the intimacy of that connection much clearer and more dramatic.

The brain is obviously involved in mental activity, and the brain is certainly a physical structure. It therefore comes as no surprise that an artificial physical structure, the computer, can perform many of the same "mental" tasks that we do. After all, many of our mental faculties are shared by animals, and no one (at least no Christian or Jew) claims that animals have a spiritual nature. Animals do have sense perception, and memory, and the capacity to make decisions. But it is important to keep firmly in mind what it is that the religious person sees as constituting the spiritual element in human beings: intellect and free will. Intellect is the power of reason, which allows us to understand ideas and to think abstractly. Free will is the power to make rational and free choices, which the mediaeval theologians defined to be "rational appetite."

Computers can do many of the things that we do, in some cases better or faster than we do them, and in other cases much worse. However, there is no evidence that they can understand anything, or that they can make free choices. So far, at any rate, the limitations of the computer correspond exactly to what the religious person would have expected. Religious teaching has, since ancient times, been very clear and very precise about where the line is between the physical and the spiritual. And everything that animals or machines can do lies clearly on one side of that line. Only when machines exhibit understanding and freedom will the expectations of the materialist be vindicated, and those of the religious believer be shown to be false.

We have come, then, to the real issue: can the human faculties of intellect and free will be understood in purely physical and mechanical terms? This is the question that I shall discuss in the remaining chapters. I shall focus on two profound and epoch-making discoveries, one in physics and one in mathematics, that seem very much to strengthen the case against the materialist view of this question and to weaken the case against the religious view.

20 | Determinism and Free Will

The Overthrow of Determinism

In the previous chapter I listed five developments in science that some see as constituting a trend in favor of materialist reductionism. Most of these I argued to be irrelevant to the religion/materialism debate, but one of them clearly is relevant. That one is the fifth item in the list, i.e., the claim that the physical world is governed by rigidly deterministic laws. As I explained, such physical determinism would be incompatible with the traditional notion of human freedom. While, conceivably, a human being could have thoughts that were free, he could not act freely in the world around him. Nor could he even express those free thoughts, since expression involves physical acts like writing or speaking.

At least since the time of Newton, and for almost three centuries thereafter, every indication coming from scientific research was that we did indeed live in such a physically deterministic world. Even before Newton, the motions of the stars and planets were observed to obey precise mathematical rules. Although the planets seemed to wander a bit, they did so in a way that could be predicted fairly well using Ptolemy's, and later Copernicus's and Kepler's, models of the solar system. In the seventeenth century Newton showed that these motions followed from a few simple equations that applied equally well to objects on Earth and in the heavens. These equations were his universal law of gravitation, and his laws of motion. As time went on more and more phenomena were found to be explicable by Newton's laws of motion: the stresses and strains of elastic media, the propagation of sound waves, the dynamics of fluids, and the flow of heat, for example. All of nature, from the tiniest atom in a gas to the largest planet, moved in precise trajectories governed by exact mathematical equations.

Thus arose the conception of the "clockwork universe." Not only massive "bodies," like atoms or planets, but rays of light also were discovered to follow exact mathematical rules, like Snell's laws of reflection and refraction. And in

175

the late eighteenth and early nineteenth century electrical and magnetic phenomena began to yield their mathematical secrets as well. Thus, by 1819, when Laplace wrote the words that I quoted in the previous chapter, a great deal was known about the character of the laws of nature. And everything that was known at that time, and for a hundred years thereafter, seemed to justify the claim that Laplace made: the state of the physical world at one time rigidly determines the state of the physical world at all future times.

If this determinism were correct, then the physical world would be "causally closed." That is, it would be sealed off from any possible influence by nonmaterial entities, if any such exist. If the human mind has any immaterial component, that component could not really be integrated with the brain and body to form a functioning whole. It would stand apart, having no effect on human behavior. Human behavior would be as rigidly determined as everything else in nature.

This issue of determinism and free will is one of the few where scientific theories have the potential of being in clear contradiction to religious doctrine; and such a contradiction really seemed to exist at the end of the nineteenth century. All that had been learned about the physical world in the preceding three centuries pointed in the same direction—toward deterministic laws of nature, while the religious believer was in the uncomfortable position of having to argue, against all scientific theory, that somewhere along the line determinism would fail.

The amazing thing is that it did fail. Completely against the expectations of the entire scientific world, determinism was overthrown in the 1920s by quantum theory. What quantum theory said was this: even given complete information about the state of a physical system at one time, its later behavior could not, in general, be predicted with certainty. Instead, the laws of physics would only permit one to calculate the relative *probabilities* of various future outcomes. Unlike "classical" (i.e., pre-quantum) physics, quantum physics is inherently probabilistic in nature.

To say that this was unanticipated would be a tremendous understatement. It was a shock, and a very unwelcome one to many physicists. Even some of the founders of quantum theory, including Einstein, Schrödinger, and de Broglie, never quite accepted its probabilistic character. Einstein derided it with the famous words, "God does not play dice." In the three-quarters of a century since quantum theory was discovered, there have been unceasing efforts to restore determinism to physics. Some of these attempts go by the names "hidden variables theories," "pilot wave theories," and the "many-worlds interpretation."[1] However, in spite of a great deal of cleverness that has been expended on these ideas, none of them has been completely successful so far. "Quantum indeterminacy" still rules. This is not to say that a future revolution in physics might not restore determinism. That is certainly conceivable. However, there is nothing in present experiments or theory which suggests that this is the direction in which physics is moving.

I will discuss quantum theory at greater length in later chapters, but it may help if I give a simple example of quantum indeterminacy here. Many nuclei and many subatomic particles are "unstable." That means that given enough time, they will "decay" or disintegrate into other nuclei or particles. For example, a "muon" will eventually decay into a group of three particles: an electron, a neutrino, and an anti-neutrino. A uranium 235 nucleus will eventually decay by "spontaneous fission" into lighter nuclei. The decay of unstable atomic nuclei is the source of the phenomenon of radioactivity. The laws of physics do not allow one to predict exactly when an unstable particle or nucleus will decay; they only allow one to assign a probability for it to happen within a certain period of time. This probability is usually stated in terms of the "half-life." Every kind of unstable particle or nucleus has a definite half-life. The half-life of muons, for example, is about one and a half millionths of a second, while that of uranium 235 is about 70 million years.

If a nucleus has a half-life of one hour, say, that means that there is a 50 percent chance of its decaying sometime within one hour. Thus, if one started out with a large number of such nuclei, after an hour only about half of them would remain, the rest having decayed. (About half, not necessarily exactly half, since it is a question of probabilities.) After two hours, only about one-quarter of them would remain, since about half of the half that remained after the first hour would have decayed in the second hour. After three hours, only one-eighth or so of the original nuclei would remain, and so on. Now, a very important thing to understand is that which nuclei decay and when is purely a matter of chance. We cannot say on the basis of any physical difference that originally existed between them why one uranium 235 nucleus decayed and another one didn't. In fact, all uranium 235 nuclei are completely indistinguishable from each other. Likewise, all muons are indistinguishable from each other. Nevertheless, in any given period of time, some will live and some will die. It is, according to quantum theory, just the luck of the draw.

I have chosen the example of the decay of particles, but in fact, according to the principles of quantum theory, all physical processes are governed by probabilistic laws. However, in most ordinary situations involving systems larger than the atomic scale, "quantum effects" and in particular quantum indeterminacy do not play a significant role and can be neglected. For such systems one is said to be in the "classical limit," and the older description in terms of deterministic "classical" laws is sufficient. From now on, I will refer to a physical process where quantum indeterminacy is important as a "quantum process." However, it should be kept in mind that all processes are, strictly speaking, governed by the principles of quantum theory.

The reason why quantum indeterminacy is not usually noticeable in big physical systems has to do, basically, with large numbers. If a system has many particles, the probabilistic uncertainties simply average out. Consider the following analogy. If one person has a 52 percent chance of voting Republican, and a 48 percent chance of voting Democratic, then there is a great deal of uncertainty in how that one person will vote. However, if there are 100 million voters, each one having precisely the same 52-48 odds, then there is virtually no uncertainty in the outcome of the election; the Republican will almost certainly get very close to 52 percent of the total vote. The expected "statistical fluctuation" in the vote total is only about 0.01 percent. That is, the Republican might well get 52.01 percent or 51.99 percent of the vote, but the chance of the Democrat winning the election would be 1 divided by a number with about ten-thousand digits.

Quantum Theory and Free Will

Not surprisingly, a number of physicists, biologists, and philosophers have made the suggestion that quantum indeterminacy has something to do with human freedom.[2] One version of this idea is that quantum indeterminacy actually explains free will. The idea is that the "freedom" of the will is simply the fact that human behavior is unpredictable, and that this unpredictability is a consequence of the random character of "quantum processes" happening in the brain. However, this version of the idea is very dubious.

In the first place, if the human mind were significantly affected by really random processes happening at the atomic level, then our behavior would be erratic and undependable. This does not correspond to what we observe in normal people, and it certainly would not have been conducive to the survival of our species. Imagine that one of our primitive forebears, in the course of sneaking up on his dinner, was suddenly impelled by a random process in his brain to jump up and down making wild cries. He probably would not have lived long enough to be a forebear.

To be subject to random mental disturbance is not freedom but a kind of slavery or even madness. Merely introducing an element of randomness into the operation of the brain, therefore, is not the way to explain free will. Furthermore, if one's only goal is to introduce unpredictability or apparent randomness into behavior, quantum indeterminacy is not required. Even a classical system that behaves in a completely deterministic way can be highly unpredictable in practice. (As I noted earlier, this is one of the characteristics of "chaotic systems.") Things that are not really random can act, for all practical purposes, as if they were random. In other words, randomness can be simulated.

For example, suppose I needed for the purpose of some game of chance a sequence of one thousand "random" digits. One thing I could do is look up in a mathematical table the first one thousand digits of the number π. The number π begins 3.141592653. These digits are not truly random, of course, since there is a precise mathematical procedure that determines what each number in the sequence is. But to a person who did not know where these digits came from they would give every appearance of being random. In fact, they would pass all the usual statistical tests for randomness. Such a sequence of digits is called "pseudo-random" by mathematicians. The upshot is that to introduce an element of mere "randomness" into the behavior of a physical system such as the brain it is not really necessary to have the true randomness of a quantum process; one can use the pseudo-randomness that is found in many deterministic non-quantum processes.

For all these reasons it does not seem that quantum theory will succeed in explaining free will. But that is not what the religious person would expect anyway. If quantum physics could explain free will, it would mean that free will was essentially nothing but a physical phenomenon. That would contradict the notion that free will is something that pertains to the "spiritual" and is therefore not completely "reducible to the merely material."

Most of those who suggest that quantum indeterminacy may have something to do with free will are saying something quite different. They do not say that quantum indeterminacy explains free will, but rather that it provides an opening for free will. Free will is conceived of as a faculty arising, at least in part, from something that is non-physical. However, in order for the non-physical to have room to operate, matter — in particular the human brain — must not be completely under the rigid control of physical cause and effect. To put it another way, true freedom implies that one actually has a choice. The alternatives one is choosing among must be real alternatives that are permitted by the laws of nature. The laws of physics, therefore, must be flexible enough that more than one outcome is possible in a particular situation: for free will to be possible, the laws of physics must have indeterminacy built into them. The key point is that quantum indeterminacy allows free will, it does not produce it.

However, some scientists and philosophers have argued that quantum theory does not even create an opening for free will. There are at least three arguments that they make:

Argument 1.   The first argument is based on the difference, emphasized earlier, between randomness and rational choice. It is claimed that quantum theory asserts that all events, to the extent that they are not determined, are governed only by chance. And if they are governed only by chance, then obviously they cannot be governed by something else, like a rational will. This argument is made, for example, by the philosopher David Chalmers in his book *The Conscious Mind*: "[T]he theory [that quantum indeterminacy allows free will] contradicts the quantum-mechanical postulate that these microscopic 'decisions' are entirely random. . . ."[3]

To understand better what Chalmers is saying, let us imagine that some decision of the human will is able to have an effect on the physical world by influencing the outcome of some "quantum process" in the brain. In particular, suppose that this quantum process is one that has two possible outcomes, and that quantum theory predicts that these outcomes have equal probability. What Chalmers is arguing is that the human "will" cannot select one of these outcomes, or even affect the odds, because that would change the probabilities from what quantum theory says they ought to be. Any influence of the will would therefore violate quantum theory.

I believe this argument to be subtly flawed and to be based on an unwarranted assumption. Specifically, Chalmers is interpreting the statements made by quantum theory in a way that is unnecessarily restrictive. If quantum theory says that two outcomes are equally probable, that can be interpreted simply as meaning that there is nothing in the physical situation itself that prefers one outcome to the other. That does not logically preclude the possibility that in certain cases something outside the physical situation may prefer one of the outcomes. If the only causes operating on the system in question are physical, and there is nothing in the physical situation that prefers one of the outcomes, then the odds will indeed be 50-50. If, however, some non-physical cause intervenes, then the odds could be different.

An analogy might be useful. If I flip a "fair" coin there is an equal chance that heads and tails will come up. That statement of probabilities is based on the fact that there is nothing about the coin itself, such as its shape or weighting, that would bias it toward one of the outcomes. But if instead of flipping the coin I deliberately place it on the table with heads up, that estimate of probabilities no longer applies, because something—my will—has influenced the situation.

Chalmers goes on to argue that if free will were to intrude itself into the physical realm it would lead to "detectable patterns" of results rather than the randomness predicted by quantum theory. This does not seem to me to be much of an argument against free will, for in practice that is just exactly what we do see when we study human behavior. There are detectable patterns in the way a human being acts, even when he is acting freely, but these patterns are neither so rigid that they can be reduced to a mechanical law, nor so erratic that they appear purely random.

Implicitly, what Chalmers is suggesting is an experimental test, which he believes would prove that free will does not affect outcomes in the physical world. To perform such a test one would have to present the same person with the same choice a large number of times. The choices would have to be the same even to the point that the person's brain, body, and environment—or at least the relevant parts of them—were known to be in exactly the same physical state prior to each choice. Using the principles of quantum theory one could then, in principle, calculate the relative probabilities for the various outcomes and compare them to the actual relative frequencies of the choices the person made. Needless to say, this is all hopelessly impossible in practice. Even supposing that some non-invasive way could be found to acquire all the relevant information about the state of a person's brain—and the human brain contains a few times 10^{25} atoms—one could never reproduce that state exactly, as required by the experiment. There is simply no way to control all the relevant variables. And even if one had all the necessary information, it would be impossible in practice to carry out the quantum-theoretical calculation of probabilities. Finally, even if all these things were possible, it is enough to point out that no such experiment has in fact ever been done, and there is no way of saying in advance what the results of it would be.

One might wonder why quantum processes are needed to produce an "opening" for free will. Could not classical, deterministic physics be interpreted in a similarly permissive way? That is, couldn't one say just as well that the precise behaviors predicted by the equations of classical physics are only to be thought of as the behaviors that would occur if no non-physical causes intervened? The answer is yes. In fact, that would be a way of allowing for free will even in a universe that had no quantum indeterminacy. However, there is a significant difference between the two cases. In the classical case, a deviation from the behaviors predicted by the physical laws due to a non-physical influence would show up as a *violation* of those laws. The laws would say that an atom should move a certain way, and it would move in a different way. For reasons that I have already indicated, it will never be possible to know if such violations of the laws of physics go on in the human brain. But the idea that they do is, to many people, rather ugly and philosophically unsatisfying.

On the other hand, in a quantum process, several alternative outcomes are truly allowed to happen by the laws of physics, and so a choice can be made without a "violation" of physical law. For example, let us suppose hypothetically that a person's brain is in a state where quantum theory says that there are only two things that he can do, A and B, and that they have an equal probability of occurring. Obviously, since the laws of physics say that he must do A or B, and give no preference to A or B, it cannot be a *violation* of those laws for him to make either choice. His freedom of choice in such a case would be unconstrained by the requirement of satisfying the laws of physics.

(There is an interesting objection that can be raised to this line of argument. Suppose that exactly the situation that we have just described is repeated one hundred times. If the person making the choices is truly free, he should be able to make the same choice—say A—all one hundred times. But the chance of this happening according to the quantum probability laws is only $(\frac{1}{2})^{100}$, or about one chance in a thousand billion billion billion. While this is not strictly a zero probability, it is prohibitively low. So it appears that the more often the same choice is presented to a person, the more his "free decisions" have to fall in line with the statistical expectations given by the quantum probability laws. It would seem, then, that quantum indeterminacy does not really provide an opening for freedom as freedom is usually understood.

One can see that this objection is really a version of the Chalmers argument that we discussed earlier. And one answer we gave there applies here too. In practice, it is not possible for a person to face exactly the same choice one hundred times, or even twice. As Heraclitus pointed out thousands of years ago, one cannot step into the same stream twice, for the stream constantly changes. If I am deciding between going to the movies tonight and staying home, it is not precisely the same as when I decided last week between going to the movies and staying home. For one thing, I am not exactly the same person physically and mentally that I was one week ago. I have different thoughts and emotions and have undergone various experiences in the meantime. The atoms in my body and brain have undergone countless changes. Moreover, I am making the choice in an altered context. Therefore, the quantum calculations that would yield the probabilities in the two situations would be very different.)

Argument 2.   A related argument that quantum indeterminacy is irrelevant to free will has been made by the mathematician and physicist Roger Penrose. In *Shadows of the Mind*, he writes, "If the 'will' could somehow influence Nature's choice of alternative that occurs [in quantum processes], then why is an experimenter not able, by the action of 'will power,' to influence the result of a quantum experiment? If this were possible, then violations of the quantum probabilities would surely be rife!"[4]

The trouble with this argument is that it is directed against a position that nobody holds. No one is suggesting that there is some general ability of the human mind to control the way all quantum processes turn out. (Such a suggestion would even have problems of self-consistency, since different people could will contradictory outcomes for the same process.) The Penrose objection might be answered simply by saying that a person's will can control only certain quantum processes that go on in that person's brain. Clearly, to say that some unpredictable events are subject to human will is not to say that all of them must be. I can believe, for example, that it is within my power to control some part of my brain so that it will lift my arm and wave it about, without believing that I can also by willpower affect the movements of the Sun, Moon, and stars.

Argument 3.   The third argument is to the effect that quantum indeterminacy is not relevant to the functioning of the brain at all. This claim is based on the assertion that those structures in the brain, specifically neurons, that are involved in mental activity are simply too large for quantum indeterminacy to play a role in them. In other words, they are large enough that the "classical limit" of quantum theory applies. This is the prevailing view of brain researchers. However, it has been challenged. Interestingly, one of the scientists who has challenged it is Roger Penrose. Penrose has argued, on the basis of certain human intellectual capacities, that quantum processes are likely to be important in understanding the brain. And he points to very small structures, called "microtubules," in the "cytoskeletons" of the brain's neurons as sites where quantum effects could potentially play an important role.[5] Sir John Eccles, a renowned neurophysiologist, has made the alternative suggestion of a certain structure called the "presynaptic vesicular grid" as a possible site for quantum effects.[6]

Whether the particular suggestions of Penrose or Eccles have any merit or not, it is certainly quite rash for anyone to assert, given the present crude state of human knowledge about the brain, that quantum processes cannot be important in explaining its behavior. In the first place, precisely because quantum effects are generally most noticeable in systems of very small size, the relevant parts of the brain are likely to be somewhat inconspicuous. Moreover, as there is no detailed theory of how the will influences the brain, brain researchers have little to guide them in their search. It is difficult to see how they would recognize the relevant structures even if they found them. It is not even clear that there would be recognizable structures.

Some humility should be induced by the fact that there are many aspects of the human mind that are not yet understood in terms of the structure of the brain. Consciousness itself is one of them. In any event, the entire history of science counsels against the overweening confidence that one finds in so much writing on this subject. Whenever hitherto inaccessible realms of phenomena are opened up to detailed investigation, dramatic discoveries follow that often baffle and surprise the scientific community. Quantum theory itself is a prime example. Absolutely no one expected, when atomic spectroscopy began in the nineteenth century, that it would lead to the strange new world of quantum physics. When the study of very low temperatures began, no one foresaw that such remarkable effects as superfluidity and superconductivity would be found. Similarly, every new tool that astrophysicists and astronomers have brought to bear in studying the universe has led to dramatic surprises, such as the existence of pulsars, quasars, galactic jets, ultrahigh-energy cosmic rays, and gamma ray bursts. Every scientist will testify to the remarkable richness and unpredictability of nature.

In the human brain one is dealing with what is agreed by all to be by far the most complex and remarkable system in nature. If there is anything that can

be predicted with confidence, it is that the investigation of the brain will produce revolutionary surprises.

How free will fits into the structure of nature remains a deep and difficult
question. However, it can be said that there is nothing in the laws of nature or
the character of physics as they exist today which is logically incompatible with
free will. After all is said and done, the fact remains that the determinism which
reigned in physical science for almost three centuries, and which seemed to
leave no place for freedom, has been overthrown. It may come back, but there
is no sign yet that it will. The great mathematician and scientist Hermann Weyl,
reflecting upon the implications of quantum theory in 1931, made the following observations, which remain valid today:

> We may say that there exists a world, causally closed and determined by pre-
> cise laws, but . . . the new insight which modern [quantum] physics affords . . .
> opens several ways of reconciling personal freedom with natural law. It would
> be premature, however, to propose a definite and complete solution of the
> problem. One of the great differences between the scientist and the impa-
> tient philosopher is that the scientist bides his time. We must await the fur-
> ther development of science, perhaps for centuries, perhaps for thousands of
> years, before we can design a true and detailed picture of the interwoven tex-
> ture of Matter, Life, and Soul. But the old classical determinism of Hobbes
> and Laplace need not oppress us longer.[7]

This was a tremendous reversal of fortunes. The religious believer of a century ago was forced to maintain in the teeth of all the physical theory of that
day that determinism was false. But as it stands now, at least, it is his expectations that have been vindicated.

Is Free Will Real?

If free will, as it is traditionally understood, is real, then scientific materialism
is certainly wrong. The reason is simple. Scientific materialism is the view that
(*a*) only matter exists, and (*b*) matter is governed by the laws of physics and nothing else. However, after four hundred years of experience, we can say something
rather definite about what a "law of physics" is. A law of physics is either deterministic or stochastic. That is, it is based on a rule or on randomness—or a com-
bination of the two.

Physics is quantitative. If human behavior is reduced to quantities, two things
only can be found. To the extent that the behavior is predictable, it falls under

some deterministic rule. To the extent that it is unpredictable, it appears to mathematical analysis as merely random. There is nothing else that a quantitative analysis can yield.

However, according to the traditional view of free will, an act is "free" only to the extent that it is *neither* random *nor* determined by rule. Like random behavior it is not predictable, but unlike random behavior it is the product of rational choice rather than chance. Free behavior is a *tertium quid*, a third kind of thing. And therefore there is no way that it can be fully explained by a mathematical theory of physics. This is almost a matter of definition.

Since free will is fatal to scientific materialism, the materialist is forced to deny its reality. This is done in two ways. Some simply assert that free will is altogether an illusion. It is a "naive" idea, a myth, and there is simply nothing in the real world that corresponds to it.

Other materialists admit that we mean something real when we talk about free will, but not what we think we mean. What free will really means, according to them, is that our choices are not coerced by causes outside our selves. In that sense a computer when it is playing chess is choosing its moves "freely." It makes its moves based upon its own internal processes. No one is reaching in from outside and overriding those processes. The computer is presented with an array of possible moves and it makes a choice among them. It is true that its program operates according to fixed, deterministic procedures, so that in a sense it cannot make any move but one in a given position. But humans also, so the argument goes, have such internal programs. It is just that, because these human programs operate largely subconsciously, we are not aware of how our choices are made or of the fact that they are really determined. Our own choices are thus opaque to us, and we attribute them to some mysterious power of the "will." In the words of Francis Crick, one of the co-discoverers of DNA, "What you're aware of is a decision, but you're not aware of what makes you do the decision. It seems free to you, but it's the result of things you are not aware of."[8] Edward O. Wilson subscribes to the same view: "The hidden preparation of mental activity gives the illusion of free will."[9] This is the theory of free will that has been advanced by Daniel Dennett,[10] among others, and has become very popular among philosophers and scientists of a materialist stripe.

Some might complain that, if freedom meant no more than this, we would not be able to hold anyone responsible for his actions. However, this is a mistake. Even if we were machines there is no reason that we could not take into account in our calculations the likely consequences of what we did, including rewards and punishments. Consequences that we were programmed to avoid would be "punishments," while those we were programmed to seek would be "rewards." A chess-playing program is programmed to avoid being checkmated, and so a move by its opponent that threatens it with that punishment affects its behavior. There are computer programs that learn from their past mistakes. To put it another way, even rats can be trained.

However, what does disappear if free will is redefined in this way is the notion that rewards and punishments can be deserved. No sensible person makes moral judgments about the behavior of animals or machines. We do not really "blame" a computer program for what it does; if anything we blame its human programmers. We do not condemn a man-eating tiger, in the moral sense, or grow indignant at the typhoid bacillus. And yet we do feel that human beings can "deserve" and that their behavior can be morally judged. We believe this precisely because we believe that human beings make free choices. When we believe that a human being is not acting freely, as in cases of truly compulsive behavior or insanity, we do not assign moral praise or blame.

The fact is that the mechanical view of free will does not correspond to our intuitions and experience. Most people believe that when faced with a choice they have real alternatives. That is, they believe that it is physically and actually possible for them to do either the one thing or the other thing. They experience this as a "power" of choice. It is not convincing to say that this is an "illusion" based simply on their ignorance of the actual unconscious processes underlying their acts. When my finger or eyelid twitches involuntarily, as happens on occasion, I am certainly not aware of the processes in my body that caused it. But even though the twitch seems unpredictable and uncaused, I do not have the experience that I am exercising a power of choice. The same is true at the mental level. Thoughts often pop into my mind "unbidden," including urges to act in certain ways. This also does not produce in me the experience of having exercised the power of choice.

What emboldens materialists to dismiss our intuitions about free will as "naive" and "unscientific" is a common view of the history of science. It is what can be called "science as debunker." Time and again, revolutionary scientific discoveries have overthrown our commonsense assumptions: It was "obvious" that Earth was at rest and the Sun moved. It was "obvious" that a thing cannot be both a wave and a particle. It was "obvious" that one can say in an absolute sense that two events happen "simultaneously." These all seemed obvious—and all of them turned out to be wrong.

In his book *Full House*, the biologist Stephen Jay Gould responded to some biting criticism he had received for some of his iconoclastic views. His critic had written, "Attentive to the adjuration of C. S. Pierce, let us not pretend to deny in our philosophy what we know in our hearts to be true." Gould gave this scathing reply:

Pierce may have been our greatest thinker, but his line in this context almost sounds scary. Nothing could be more antithetical to intellectual reform than

an appeal *against* thoughtful scrutiny of our most hidebound mental habits—
notions so 'obviously' true that we stopped thinking about them generations
ago, and moved them into our hearts and bosoms. Please do not forget that
the sun really does rise in the east, move through the sky each day, and set
in the west. What knowledge could be more visceral than the earth's central stability and the sun's subordinate motion?[11]

There is some truth to what Gould says, of course. Thoughtful scrutiny of hidebound mental habits is a good thing. Nevertheless, what is obvious to our intuition has to be accorded great weight. It cannot simply be set aside. And the
conventional examples taken from scientific history to prove the contrary really
fail to do so.

First of all, most of the "obvious" statements found to be false by science were
not simply based on "illusions" in the sense of something being seen that was
not there. Consider Gould's example of the motion of the earth. Everyone knew,
including Galileo's scientific opponents and the head of the Roman Inquisition,
that one's own motion can cause the apparent motion of other objects. This was
the point of the Copernican theory, and it was not lost upon Galileo's contemporaries. If they argued from the "obvious" it was not simply from the obvious
fact that the Sun appeared to move.

One of the arguments that was used against Galileo was that if Earth were
spinning on its axis we would feel the motion. Now, in point of fact, the physical
intuition upon which this argument was based is completely correct. A rotating object is indeed subject to centrifugal forces; and that is why when we spin
around we can feel that we are moving. The problem is that the rate at which
Earth spins—its "angular velocity"—is far too small for the resulting forces to be
perceptible to us, though they can be measured by sensitive instruments.

Or consider the intuitions about space and time that were overthrown by the
theory of relativity. These intuitions also were not just blunders or "illusions."
Objects that move slowly do in fact behave as these intuitions tell us. Our intuitions about motion—about "kinematics"—are Newtonian, and Newtonian
physics is not simply "wrong." It would be more accurate to say that it is right
"up to a point." Newtonian kinematics is the one and only correct "limit" of Einsteinian kinematics when velocities are small compared to the speed of light,
which is the regime in which we live our daily lives.

Thus, these naive intuitions about physics, rather than being illusions, correctly correspond to real features of the world around us. In most cases they are
approximations to the truth that are valid if they are not extrapolated too far from
the range of experience on which they are based.

The same is true of some of our mathematical intuitions. For example, we
feel intuitively that a group of objects will have fewer things remaining in it if

we take away some of them. However, it turns out that this is not always so. Specifically, it is not always so of infinite sets of objects, as was shown by the brilliant mathematician Georg Cantor about a century ago. But, again, our intuition about numbers was not just mistaken; it was a genuine mathematical insight. We would only be wrong if we jumped to the conclusion that what is true in a familiar setting would also be true in the bizarre realm of infinite numbers, of which we have no experience in everyday life.

But there is an even more fundamental point. The naive ideas that were overthrown by the progress of physics had to do with the nature of things outside our minds. These are things we only know indirectly through our sensations and mental experiences of them. To construct an accurate account of the external world from our experiences and sensations requires a very complicated set of inferences. On the other hand, our own mental acts and powers are experienced by us directly, not through the mediation of something else. As the physicist Andrei Linde observed, "Our knowledge of the world begins not with matter but with perception. I know for sure that my pain exists, my 'green' exists, my 'sweet' exists. I do not need any proof of their existence, because these events are a part of me; everything else is a theory."[12]

In fact, the very existence of our own brains is an inference. Even if a person could see his own brain or touch it, its existence as a real object with definite properties would remain an inference drawn from his sensory experiences. However, we do not have to infer the existence of our minds. We experience them directly in the process of using them. We do not infer the existence *of* our minds, rather we infer the existence of everything else *with* our minds. To put it another way: the brain does not infer the existence of the mind, the mind infers the existence of the brain. We may doubt our inferences about external things, but it is madness to doubt the reality of the mind itself and the reality of its basic powers.

The point may be made clearer by considering the difference between light and the sensation of light. Before physicists could really understand the nature of light they had to do a lot of experimentation and develop some fancy mathematics. It turned out that light consists of electromagnetic waves. It would not be at all surprising if someone's ideas about light were naive. However, it is one thing to tell a person that he is mistaken about the nature of light, and something else entirely (something quite ridiculous) to tell him that he is not experiencing a sensation of brightness or color when he is. He may be mistaken about what produced the sensation, but not about the fact that he is having a sensation.

In the same way, it is somewhat ridiculous for a person to argue on the basis of rudimentary brain anatomy or speculative ideas in physics that we do not exercise free will. Dr. Johnson, the great eighteenth-century lexicographer and man of letters, who was as famous for his massive common sense as for his literary gifts, made this point in his characteristically incisive way. In the course of a discussion on free will he posed the following question to Boswell: "If a man should give me arguments that I do not see, though I could not answer them, should I believe that I do not see?"[13]

On another occasion when Boswell tried to draw him into a debate about free will, Johnson declared flatly, "Sir, we *know* our will is free, and *there's* an end on't!"—an assertion that was certainly dogmatic, but not at all unscientific.[14] There is a certain degree to which we must trust our experiences if we are to do any rational thinking at all, including scientific thinking. There are many things we must not regard as illusions. Among these are the external world itself; our conviction that there is an objective truth about that world; our own rationality; our belief that certain kinds of rational arguments are objectively valid; our own existence; and, in a general way, our powers of sense and memory. Our senses or memory may occasionally play tricks on us, but there must be a presumption in their favor, since if everything we sense and everything we remember, even about what happened one second ago, could be illusory, then we would have no basis for knowing anything whatsoever.

Those who think being "scientific" means being ready at the drop of a hat to debunk anything that seems obvious, or being willing to disregard all that is intuitively known, are walking a foolish and treacherous path. It is a path that leads in the end to pure skepticism, a skepticism not only about religious doctrines but about everything, including science. There was a time when religious skeptics proudly called themselves "free thinkers." It is ironic that the modern materialist skeptic disbelieves even in the reality of his own freedom, both moral and intellectual.

21 | Can Matter "Understand"?

Having discussed free will, let us now turn to the other faculty of human beings that is claimed to distinguish us from the lower animals and from machines, namely intellect or reason. The crucial question is whether the human intellect is something that can be explained in purely physical and mechanistic terms, or whether it points to the existence of a reality that goes beyond the physical. As we shall see in the next chapters, a revolutionary discovery made in the early part of the twentieth century casts considerable light on this question. This discovery, unlike those we have discussed up to this point, is in the field of mathematics rather than physics. It is a theorem proven in 1931 by the great mathematical logician Kurt Gödel. On the basis of this theorem it has been argued, notably by the philosopher John Lucas and the mathematician and physicist Roger Penrose, that all of the powers of the human intellect cannot be accounted for on the supposition that the human mind is merely a computer in action. Before getting into the details of the Lucas-Penrose argument and the technicalities of Gödel's Theorem, it will be helpful to set the stage by approaching the question from a more philosophical point of view.

The first thing that we must do is be a little clearer about what we mean when we speak of the intellect. Without attempting to be comprehensive, we can certainly say that among the powers of the human intellect are conceptual understanding and rational judgment. The former is the ability to understand the meanings of abstract concepts and of the propositions that contain them, while the second is the ability to judge the adequacy of these concepts and the truth of these propositions. These are the human abilities with which we will be chiefly concerned in the chapters to come.

A point that is important to emphasize is that the powers of intellect and free will, even though we are discussing them separately, are intimately related. Both are aspects of human rationality. What makes will free, as we noted in the previous chapter, is that it is rational. We can have reasons for our decisions. We can deliberate about them. They can be based upon judgments about what is true and false or right and wrong. And, by the same token, our intellects if they are truly rational must be free rather than controlled by merely physical causes. Human freedom and human rationality stand or fall together.

ABSTRACT UNDERSTANDING

When we say that the intellect has the power to understand abstract concepts, we have to explain what we mean by the word *abstract*. This is perhaps best done with a simple example. If you think about a particular woman, you are not thinking abstractly, but if you think about "womanliness," you are. Abstract terms, such as *womanliness, truth, impossibility, virtue*, or *circularity*, are what philosophers call "universals." The word *circularity*, for instance, is a universal because it does not refer to this or that circular object—this dinner plate or that wagon wheel, say—but to all real or possible circular objects. Indeed, it can be understood apart from any concrete example of circularity. When we think of a universal apart from any particular object, we are engaged in truly abstract thought.

Abstract thought has, therefore, in a sense, an unlimited reach. It transcends what is here and now and the particularities of specific objects to embrace a concept that is infinite in scope. For this reason, the philosopher Mortimer Adler,[1] following a philosophical tradition going back at least to Aristotle,[2] has argued that nothing which is merely material can engage in abstract thought. It is true that a material system can exemplify—or in philosophical jargon "instantiate"— a universal, the way a dinner plate exemplifies circularity. But it cannot contain the whole abstract meaning of a universal. For example, a dinner plate is composed of certain materials, exists at a certain time, and has a certain size, position, and orientation in the three-dimensional physical space of our world. The concept "circularity" has no such limitations. It applies to circles of any size, position, and orientation. Indeed, it applies even to circles in numbers of dimensions that cannot be "instantiated" in our physical world. (The human mind cannot visualize more than three dimensions, but the human intellect is nevertheless able to think abstractly about mathematical entities, like circles and spheres, in any number of dimensions. For example, any mathematics graduate student could easily prove that the volume of a four-dimensional sphere is $\frac{\pi^2}{4}$ times its radius to the fourth power.)

The argument, then, is that because our brain is a finite material system, it cannot encompass within itself the whole meaning of an abstract concept. It may contain images that illustrate abstract concepts. It may even have words or symbols stored in it that "stand for" abstract concepts. But the full universal meaning of an abstract concept cannot be inscribed in it. There must, therefore, according to Adler, be some non-material component to our minds that enables us to think abstractly.

Adler also maintains that there is no scientific evidence that any animal other than human beings can understand universals. He admits that there are some facts that appear at first sight to contradict this. For example, even some species of fish can distinguish between a square object and a circular object. However, this is not an example of true abstract thinking, according to Adler, but rather of what he calls "perceptual abstraction." These fish can only recognize a circle when presented with a circular object. In other words, the "abstraction" is tied closely to a perceptual act. In contrast, human beings can engage in what Adler calls "conceptual abstraction"; they can think about circularity in general, apart from any perceived object. They can relate the concepts circle and circularity to other concepts, or make them mathematically precise, or even prove theorems about them.

At this point, a materialist would be tempted to object that computers can prove theorems also, and therefore (since computers are material objects) it must be that Adler is wrong. But this objection raises the question of whether a computer really "understands" what it is doing. It manipulates symbols or numbers, and those symbols or numbers mean something to the human programmer. But do they mean anything to the computer itself? Does it know that the string of symbols it prints out refers to circles rather than to rocks or trees? Does it know that it is saying anything meaningful at all? The symbols or "bits" that it manipulates stand (in somebody's mind—but not the computer's) for concepts, but the symbols and bits are not, themselves, concepts.

Many materialists would reply that a computer can understand, because "understanding" information means no more than being able to put that information to appropriate use, to act in appropriate ways on the basis of that information. They would say that if a robot uses information from sensors to avoid bumping into a table, it "understands" that the table is there. However, it would seem that understanding is something more than this. There are many things that we understand that have no particular relevance to our behavior. We have insights that we cannot possibly put to practical use. If we understand something about spheres in six dimensions, what is the appropriate behavior that follows from that understanding? After all, we do not live in six dimensions. An aristocratic lady once asked the famous nineteenth-century Irish mathematician William Rowan Hamilton, who had invented "quaternions" (a kind of number), what quaternions were useful for. He replied, "Madam, quaternions are very useful for solving problems that involve quaternions."

The idea that a computer "understands" if it can make use of information leads to rather bizarre conclusions. An ordinary door lock has "information" mechanically encoded within it that allows it to distinguish a key of the right shape from keys of other shapes. It uses that information to react in an appropriate way when the right key is inserted into it and turned: the lock mechanism pulls back the bolt and allows the door to open. Does the lock understand anything? Most sensible people would say not. The lock does not understand shapes any more than the fish understands shapes. Neither of them can understand a universal concept. However, many materialists do believe that even very simple non-living physical systems have mental attributes. For example, the man who invented the term *artificial intelligence*, John McCarthy, has written that "machines as simple as thermostats can be said to have beliefs."[3]

WHAT ARE ABSTRACT IDEAS?

In talking about thermostats, McCarthy was not choosing an example at random. A thermostat can be thought of as a very simple brain. For, just as an animal's brain receives information about the world around it from sensory organs, a thermostat has a sensor which tells it what the temperature is in some particular locale, such as the living room of a house. And just as an animal's brain controls a body, telling it how to react to its environment, a thermostat controls some apparatus, usually a heating or cooling device. We could say, then, that a thermostat "senses" one feature of its environment and responds to it. And we therefore could, in some very broad sense, attribute to thermostats "sensation," "perception," and even "cognition." However, it remains the case that a thermostat cannot understand universal concepts or abstract ideas. To put it succinctly, a thermostat does not understand thermodynamics.

Any sane materialist would concede this, of course, but would explain it by saying that a thermostat is simply much too elementary a brain to have higher mental functions like abstract thought. What makes it possible for human beings to think abstractly, in the materialist's view, is the enormous complexity of our brain's neural circuitry.

While this view may sound plausible at first, it leads to conclusions that would be uncomfortable, I think, to most physical scientists, even, I dare say, to many scientists of a materialist bent. To see the difficulty, let us consider the difference between perceiving a physical object, like a tree, and thinking about an abstract idea like those encountered in mathematics, such as the number π.

The impression produced in our minds by a tree is made up of various sensations, both present and remembered, like the roughness of bark, or the rustling of leaves, or the filtering of light through the foliage, together with many associations to ideas related to trees, such as shade, forests, timber, wood, songbirds, and so on. We can easily imagine that all of these mental impressions are just a complicated set of responses of our nervous system and brain to the external physical reality we call a tree.

Thinking about an abstract idea like the number π is in some respects similar to this. Certainly, many images and associations crowd around the idea of the number π in the mind of the mathematician or physicist. However, in an obvious and important way there is a great difference between thinking about a tree and thinking about π. A tree is a physical object, and π isn't. The tree presents no problem for the hard-nosed materialist who believes that only matter exists, for a tree is made of matter. But π is not made of matter. So what is the materialist to make of it, or indeed of any other number, or mathematical concept?

An obvious answer is that a number, while not itself a material thing, is an aspect of material things. While the number 4, for example, is not made of matter, a 4-sided table is, and 4 rocks are, and a 4-footed animal is. The number 4, then, might be considered a property of material objects. When we think about "4," we are really thinking about things or groups of things in the real physical world that in some way have four-ness about them. While plausible for small "counting numbers" like 4, this point of view runs into serious difficulties when it comes to other kinds of numbers like π. One can have 4 cows, but one cannot have π cows; and one can have a 4-sided table, but not a π-sided table. Of course, in a sense, one can have a π-sided table: a circular table whose diameter is one meter will have a circumference of π meters. So can π perhaps really be thought of as simply a property of material objects, and specifically of objects that are circular?

This is a very common idea. In fact, if you ask "the man in the street" what π is, this is the answer you are likely to get. But it really will not do. One problem with it is simply that there are no exactly circular objects in the physical world; and for an object that is not exactly circular, the ratio of circumference to diameter is not π (except for very special shapes that are as unlikely to exist in nature as exact circles). It might be close to π, but close to π is not π, at least not the mathematician's π. Nevertheless, π does at least have some connection to shapes that we see approximated in the physical world. However, most numbers do not have even this link to the world of matter. Consider, for example, the number 0.101100111000111100001111 .... (The pattern is clear: one 1, one 0, two 1's, two 0's, three 1's, three 0's, etc.) This is a definite well-defined number, but it has no connection to any shape or figure that one will find in the physical world. Nor will most other numbers, such as the 17th root of 93.

We are left with a problem. If numbers and other mathematical concepts (unlike trees) are neither material things nor even aspects or properties of material things, then what are they? The most reasonable answer seems to be that they are mental things, things that exist in minds. Mathematics is a mental activity. Most schools of thought in the field called "the philosophy of mathematics" adopt some version of this view. But this raises the question of what a "mind" is and what "mental things" are.

To the non-materialist, minds and the ideas they contain can be real without being entirely reducible to matter or to the behavior of matter. To the materialist, however, there can be nothing to our minds besides the operations of our central nervous systems. In the memorable words of Sir Francis Crick, "you are nothing but a pack of neurons."[4] Consequently, to a materialist, it follows that "an explanation of the mind . . . must ultimately be an explanation in terms of the way neurons function," to quote Sir John Maddox, the former editor of the scientific journal *Nature*.[5] Now, if we say that abstract concepts, such as the number π, exist only in minds, and if we also say, with the materialist, that minds are only the functioning of neurons, then we are left in the strange position of saying that abstract concepts are in themselves nothing but patterns of neurons firing in brains. Not, mind you, merely that our neurons fire when we think about or understand these concepts, or that the firing of neurons plays an essential role in our thought processes, but that the abstract concepts about which we are thinking are *in themselves* certain patterns of neurons firing in the brain, and *nothing but* that. That is why one finds in a recent article the statement, "Numbers are . . . neurological creations, artifacts of the way the brain parses the world."[6] The author of that statement was summarizing the views of a "cognitive scientist" who had written a book subtitled "How the Mind Creates Mathematics" (by which he really meant, of course, "how the human central nervous system creates mathematics").[7]

To the consistent materialist, then, the number π is a pattern of discharges of nerve cells. It has no more status, therefore, than a toothache or the taste of strawberries. This is a notion that many people who deal extensively with abstract mathematics would have a hard time accepting. The number π appears to the mathematician as something more than a sensation or a neurological artifact. It is not some private and incommunicable experience, like a toothache; it is a precise, definite, and hard-edged *concept* with logical relationships to other equally precise concepts. It is something that can be calculated with arbitrary precision; for example, to ten figures it is 3.141592653 (and it has been calculated to 50 billion figures). It has remarkable and surprising properties, which the mathematician feels that he is discovering rather than generating neurologically.

To take just a few examples, the sum of the infinite series of fractions $1 - \frac{1}{3} + \frac{1}{5} - \frac{1}{7} + \frac{1}{9} - \frac{1}{11} + \ldots$ is exactly $\frac{\pi}{4}$; the sum of the infinite series $1 + (\frac{1}{2})^2 + (\frac{1}{3})^2 + (\frac{1}{4})^2 + \ldots$ is exactly $\frac{\pi^2}{6}$, and the natural logarithm of -1 is $i\pi$, where i is the square root of -1. But what are all these precise and beautiful mathematical statements? According to the consistent materialist they too are "neurological creations." Not only π itself, but the statement $1 - \frac{1}{3} + \frac{1}{5} - \frac{1}{7} + \frac{1}{9} - \frac{1}{11} + \ldots = \frac{\pi}{4}$ is no more, ultimately, than a pattern of neurons firing in someone's brain. The neurons firing in someone's brain in a certain way may lead him to write certain figures on a piece of paper or blackboard (like the formula "$1 - \frac{1}{3} + \frac{1}{5} - \frac{1}{7} + \frac{1}{9} - \frac{1}{11} + \ldots = \frac{\pi}{4}$"), and those shapes on paper or blackboard may in turn stimulate the neurons in someone else's brain to fire in certain patterns. But whether it is patterns of ink on paper, of chalk on blackboard, or of neurons firing in brains, concepts, to the consistent materialist, are just patterns that exist in some material system.

Some materialists might be tempted to explain their position by saying that these physical patterns on paper or in the brain "mean" something. However, to say that a pattern means something is to say that it stands for some ideas: meanings are ideas that are understood by minds. And from the materialist's point of view to say that a "meaning" is being "understood" by a "mind" is ultimately to say no more than that some pattern of electrical impulses is occurring in a brain. The materialist cannot go beyond patterns to the "meanings" of the patterns, because meanings themselves are ultimately nothing but patterns in brains.

I have used the number π as an example, but I could have used any other abstract concept, whether mathematical or not. However, since mathematics is the realm of the purest abstract thought, and is also the language of physical science, it is particularly relevant to our discussion. For if we reduce mathematical ideas to neurons firing we reduce all of scientific thought to neurons firing. What is the theory of relativity? What is quantum theory? What is the Schrödinger equation? What are Maxwell's equations of electromagnetism? Just neurons firing? What are the statements theoretical physicists make, like "Observables are hermitian operators acting in a Hilbert space," or "All Cauchy surfaces for a space-time are diffeomorphic," or "Spontaneously broken gauge theories are renormalizable"? Nothing but patterns of nerve impulses? Squiggles on a page? It is the absurdity of this kind of conclusion that was the basis of the incisive critique of materialism made by Karl Popper, the eminent philosopher of science, especially in his later works.[8]

Most people do not like mathematics and physics much, and perhaps would be just as happy to think of them as some quirky phenomena occurring in the nervous systems of a small number of peculiar people. (Not surprisingly, most mathematicians and theoretical physicists do not share this view.) But it is not just the concepts of mathematics and physics that are at stake here; all concepts are at stake, including those of biology, neuroscience, and indeed the concepts of the cognitive scientist I just mentioned who thinks that numbers are "neurological creations." Cognitive scientists talk about neurons, for example. But "neuron" itself is an abstract concept that arose from the researches of biologists. For the materialist, then, even this concept of "neuron" is nothing but a neurological creation; it also is a pattern of neurons firing in someone's brain. If this sounds like a vicious circle, it is. We explain certain biological phenomena using the abstract concept "neuron," and then we proceed to explain the abstract concept "neuron" as a biological phenomenon—indeed, a biological phenomenon produced by the activity of neurons. What we are observing here is the snake of theory eating its own tail, or rather its own head. The very theory which says that theories are neurons firing is itself naught but neurons firing.

This is an example of what G. K. Chesterton called "the suicide of thought."[9] All of human understanding, including all of scientific understanding, is reduced to the status of electrochemical processes in an organ of the body of a certain mammal. In the words of a recent *Newsweek* article, "Thoughts . . . are not mere will-o'-the-wisps, ephemera with no physicality. They are, instead, electrical signals."[10]

Why should anyone believe the materialist, then? If ideas are just patterns of nerve impulses, then how can one say that any idea (including the idea of materialism itself) is superior to any other? One pattern of nerve impulses cannot be truer or less true than another pattern, any more than a toothache can be truer or less true than another toothache.

TRUTH

The intellect has not only the power of abstract understanding but also the power of judging the truth and falsehood of propositions. It has been argued by philosophers since antiquity that this power goes beyond the capacities of purely physical or mechanistic systems. There are many grounds for this conclusion; one of the most basic has to do with what might be called "openness to truth." Human "openness to truth" was, in fact, referred to in one of the passages quoted in chapter 19 as a sign of our "spiritual" nature.

The question is this: If my thoughts follow a path set out for them by the laws that govern matter, how does "truth" enter into the picture? In the final analysis, my thoughts are not "reasonable" or "unreasonable," they are just the thoughts that I *must* have given the way the molecular motions in my brain and the rest of the world have happened to play out. As the mathematician Hermann Weyl put it, "[there must be] freedom in the theoretical acts of affirmation and negation: When I reason that 2 + 2 = 4, this actual judgment is not forced upon me through blind natural causality (a view which would eliminate thinking as an act for which one can be held answerable) but something purely spiritual enters in."[11] He went on to explain that thought that is free, and therefore rational, must not be entirely determined by physical factors (in which case it would be "groundless" and "blind") but must be "open" to meaning and truth.

Writing in the same year (1932), the biologist J. B. S. Haldane argued, "If materialism is true, it seems to me that we cannot know that it is true. If my opinions are the result of the chemical processes going on in my brain, they are determined by chemistry, not the laws of logic."[12] In *Orthodoxy*, his brilliant defense of Christianity written in 1908, G. K. Chesterton noted that the materialist skeptic must sooner or later ask, "Why should anything go right; even observation and deduction? Why should not good logic be as misleading as bad logic? They are both movements in the brain of a bewildered ape."[13] Recently,

Stephen Hawking worried about the same issue in connection with the "theory of everything," which many physicists are seeking.[14] A theory of physics that explained everything would also have to explain why some people believed it and some people did not. Their belief (or disbelief) in the theory, then, would be the result of inevitable physical processes in their brains rather than being a result of the validity of the arguments made in behalf of the theory itself. This argument is very ancient. For example, Epicurus wrote, "He who says all things happen of necessity cannot criticize another who says that not all things happen of necessity. For he has to admit that the assertion also happens of necessity."[15] (Of course, the necessitarian could always make the clever comeback that not only can he criticize, but he must criticize—*of necessity*. However, Epicurus's point is that the criticism would not then be a rational one.)

Haldane recanted his argument in 1954 because of the development of the computer.[16] He was impressed by the fact that although a computer is made only of matter and obeys the laws of physics it is nonetheless capable of operating in accordance with truth. However, Haldane was wrong to recant, since the example of the computer does not really resolve the question that he originally raised. It is true that a calculating device can print out $2 + 2 = 4$, or some equivalent formula, and in that sense can operate in accordance with the truth. However, its ability to do so is not to be sought simply in the laws of physics that it obeys. A calculating device (also obeying, of course, the laws of physics) could just as easily be produced that printed out $2 + 2 = 17$. In fact, it would be even easier to build a device that printed out complete gibberish. The reason that most calculating devices do operate in a manner consistent with logic and mathematical truth is that they were *programmed* to do so. That is, they have built into them a precise set of instructions that tells them exactly what to do at every step. These programs are the products of human minds. More precisely, the acts of understanding that lie behind these programs took place in human intellects. Rather than illustrating, therefore, how an automatic device can give rise to intellect, artificial computers merely show that an intellect can give rise to a device. Not only do the design and programming of these devices occur as the result of human acts of understanding, but the meaning of their outputs can only be apprehended by human acts of understanding, not by the machines themselves. (These outputs can indeed be *used* by other machines, but only by machines designed to be able to do so by human intelligence.)

The mystique of the high-performance computer can obscure what is really going on. Let us therefore consider instead a humbler device, a vending machine. A vending machine contains a simple computing device that enables it to make change correctly. In spite of this, we do not attribute intelligence to vending machines. On the other hand, we might well attribute a great deal of intelligence to a child who was able to figure out for himself or herself how to make change correctly. What is the difference? The difference is that the child understands something and the vending machine does not. Of course, the vending machine could not do what it does without intelligence being involved somewhere along the line. At some point there was an understanding of numbers and the operations of arithmetic; there was an understanding of how to perform these operations in a routine fashion; and finally there was an understanding of how to build a machine to carry out these routine steps automatically. All of these acts of understanding took place in human intellects, not in vending machines. The point at which any task has become routinized so that it no longer requires acts of understanding is the point at which it can be done by a machine which lacks intellect.

The question raised by progress in developing computer hardware and computer programs is whether a sufficiently advanced computer could have genuine intellect, and whether in fact the human intellect can be explained as being the performance of an enormously powerful biological computer, the brain. When we "understand" abstract ideas, is no more going on than that our brains are following some canned instructions, some very complex but routine procedure? That is the issue that we will examine in the next chapters. For now let us suppose (contrary to what we shall argue later) that the human intellect can indeed be explained as the product of a sophisticated computer program.

A number of difficult questions then arise. For example, how is it that we know, as sometimes at least we do know, that our thought processes are reasonable and consistent? As we shall see later, no program (except a very trivial one) that operates consistently is able to prove that it does so. If we were merely machines, then, we could not be aware of our own consistency. Moreover, if our brains have in fact been programmed so that they can think consistently and reasonably, how did that happen?

If an electronic computer operates in a correct way, it is always because some human beings programmed it to do so. But who programmed those human beings to operate correctly so that they could impart that correctness to the computer? The only answer available to the materialist (and the one suggested by Hawking, among others) is natural selection. Natural selection programmed human beings to think in such a manner that our thoughts correspond in some way to reality. Obviously, an organism that could not think straight would be at a disadvantage in the struggle for survival. This answer is appealing at first sight, but it is really not adequate.

In the first place, the human mind is capable not only of dealing with the kinds of problems that our primitive forebears faced, like how to escape from a predator, or how to make a rude shelter; it is also capable of doing an enormous variety of things that have no conceivable application to survival in the wild, like playing chess, proving theorems in non-Euclidean geometry, doing research

in nuclear physics, designing jet aircraft, or composing violin concertos. If the human mind is nothing but a computer program, this versatility is quite mysterious. A program that was designed to play chess would not, in general, be able to do other things, even to play other games, like Parcheesi. A computer program that could play many types of games would be much more complicated than a program that could only play one. Similarly, a program (assuming one were possible) that could replicate human abilities in every field of activity would be far more complicated in structure than a program that had only the ability to do the things that cavemen did in order to survive. How can one possibly explain that natural selection gave us a program than was vastly more sophisticated than was required for survival?

This kind of argument (among others) has led Roger Penrose to challenge the idea that the human intellect operates simply as a computer program. He argues that what enabled our forebears to survive was not simply a complicated program, but something which no mere program can have, namely the capacity to "understand"—that is, intellect.[17] And since the capacity to understand is a very general kind of thing, it allows human beings to perform well in a wide variety of activities. I shall have more to say about Penrose's ideas in the next chapter.

There is another problem with the idea that the human mind is merely a computer programmed by natural selection, and this has to do with two remarkable abilities which the human mind possesses: the ability to attain *certainty* about some truths, and the ability to recognize that some truths are true *of necessity*. (These may seem like the same thing, but they are not. I am certain that my first name is Stephen, but it could have been something else, so that is not a "necessary truth." On the other hand, it is necessarily true that $147 \times 163 = 23,961$, in the sense that it could not have been anything else, and must be so in any possible universe, but someone who is not adept at multiplying large numbers might nevertheless be in a state of uncertainty about it.)

There are two aspects to this problem. In the first place, a creature's "evolutionary success" (that is, its success in surviving to reproduce and in ensuring the survival of its offspring) does not require that it know things with absolute certainty or that it recognize truths as necessary ones. It is quite enough for it to have knowledge which is reliable for practical purposes and which is known to be generally true in the circumstances that it has to face. It does not have to know with absolute certainty that this branch will support its weight, that this fruit is not poisonous, or that fire will burn it, in order to survive. It is quite enough for it to be 99.99 percent sure or even 90 percent sure. Nor is it of any use to it to understand that the statement "$2 + 2 = 4$" is true of necessity, and therefore true in any possible world. It would be just as good to know that $2 + 2$ can be trusted to come out 4 in its own situation.

The second aspect of the problem is that even if it were helpful to their survival for human beings to have absolute certainty in some matters, or to realize that some things are true of necessity, there seems to be no way that natural selection could possibly have programmed us to have that kind of knowledge. Natural selection is based ultimately on trial and error. Various designs are tried out, including various designs of brain hardware and brain software, and those that give the best results on average tend to lead to more numerous offspring. However, trial and error cannot produce certainty. Nor, obviously, can it lead to conclusions about what is necessarily true.

In a recent book about the philosophical implications of modern science the author asked, in all seriousness, "Is it so inconceivable that a reality *could* exist in which 317 is not a prime number?"[18] The answer is, quite simply, yes. It is totally inconceivable, indeed absurd. I do not know of any scientist or mathematician who would admit any possibility of doubt about this. Not only is 317 prime here and now, it is indubitably prime in galaxies too remote to be seen with the most powerful telescopes. It was prime a billion years ago and will be prime a billion years hence. It would have to be prime in any other possible universe. However, there is no way that the processes of natural selection that operated upon our forebears could have had access to information about what will be true in a billion years, or in remote galaxies, or in other possible universes. How, then, can those physical processes have taught us these truths, or fashioned us so that we could recognize them?

It is important to be clear about what the issue is here. The issue is not how we came to be able to do arithmetic and figure out whether 317 is a prime number. Having the abilities that enable us to figure out the rules that will give correct answers to arithmetical problems may indeed have advantages for survival. One could even imagine that by evolutionary trial and error the right circuitry was "hard-wired" into our brains to do arithmetic correctly. And, therefore, assuming that "317 is a prime" happens actually to be a necessary truth, it is not surprising that evolution allows us to arrive at conclusions which in fact happen to be necessary truths. The question, however, is this: How do we *recognize* that necessity? How and why did natural selection equip us, not merely to say that 317 is prime, but to recognize of that truth that it is true of necessity?[19]

(An aside to philosophers: It may appear that I am implicitly endorsing the Kantian idea that mathematical statements are synthetic *a priori* truths. I am not. Even if one adopts the view, say, that mathematics is reducible to logic, precisely the same kind of question arises with respect to evolution: How was natural selection able to produce in us the knowledge that the laws of logic are universally valid and admit of no exceptions?)

The reaction of many materialists to such an argument, I dare say, would be to suggest that human beings are not actually capable of achieving certainty about anything, or knowledge of the necessity of truths. In their view, all we can ever claim to have is knowledge that has a high probability of being right. When we say we are "certain," we are, according to this view, only expressing a strong feeling of confidence in what we are saying. And when we say something is "true by necessity," we just mean that we have not been able to imagine a contrary situation. Absolute certainty, according to many materialists, is a chimera.

This skeptical position has some superficial plausibility. After all, we have all had the experience of claiming to be certain about something only to find out later that we were mistaken. In spite of its initial plausibility, however, this account of what we mean by "being certain" is, in my view, simplistic and untenable.

Consider the two statements, "The Sun will come up tomorrow" and "317 is a prime number." I have great confidence in the truth of both. But they are radically different types of statement, in which I have radically different types of confidence. I admit that it is overwhelmingly probable that the Sun will come up tomorrow, but I do not believe that it is absolutely certain. It is quite conceivable that the Sun will not come up tomorrow, and in fact there are scenarios, not excluded by anything that we know about the laws of nature, in which the Sun would not come up tomorrow.

To take the most exotic such scenario, we could be in what particle physicists call a "false vacuum state." That is, just as some radioactive nuclei with very long half-lives appear to be stable, but actually have a small chance of disintegrating suddenly and without prior warning, so the state of matter of our world may actually be unstable in the same way. It is possible that a large bubble of "true vacuum" — that is, a state of lower energy — will suddenly appear in our vicinity by a "quantum fluctuation." If it does, it will expand at nearly the speed of light and destroy all in its path. We would never know what hit us. The Sun would not come up tomorrow, because the Sun would have ceased to exist. (No particle physicist loses even a moment's sleep over this possibility, but none would claim that it is absolutely ruled out either.)

There are less exotic possibilities that also are not excluded by what is presently known about physics and that would prevent the next sunrise. And, aside from natural catastrophes, there is always the logical possibility of a miracle. Earth might miraculously stop rotating on its axis or the Sun might miraculously disappear. As the philosopher David Hume pointed out, one cannot rigorously deduce what will happen in the future from what has happened in the past. Therefore, the skeptical materialist is right about this case: when we say we are "certain" that the Sun will come up tomorrow, what we really mean is that we have an extremely high degree of confidence that it will.

However, it is far otherwise with "317 is a prime number." No scientific phenomenon, however exotic, can make 317 not be a prime. The mediaeval theologians would have said that even the omnipotence of God could not do that.[20] This is not just a question of something that is very highly probable, but of something truly certain. Someone might object that 317 is a fairly large number, and that the calculations which show it to be prime are too complicated to allow him to be certain about them. However, one can always take a case where this is not an issue, like $1 \neq 0$ (1 does not equal 0). I think most people would admit to knowing this with certainty, and not simply to having a lot of confidence in it.

Moreover, the idea that our certainty about such things as "317 is prime" is simply a sort of gambler's confidence, a confidence born of practical experience, simply does not bear careful scrutiny. I have seen that the Sun came up about fifteen thousand times without fail; whereas in the last fifteen thousand arithmetical calculations I have done, I have not always gotten consistent answers. In fact, on many occasions I have not. If anything, then, my confidence that the Sun will come up tomorrow ought to be *greater* than my confidence in the consistency of arithmetic. I can make the argument even stronger. It is clear, from the absence of excited testimony to the contrary, that the Sun has come up every day without fail for the last several thousand years—about a million consecutive times.

However, our confidence in arithmetic is in fact stronger than our confidence in the Sun coming up. Why? Is it based on some logical analysis? But that just raises the question of how it is that we have the confidence we do in the consistency of logic.

Even if one were to concede that our certainty is never absolute (which I am not prepared to do), it would still remain the case that we have a certainty about some kinds of truths that far exceeds what we can derive from trial and error, and which it is very hard to explain as arising from natural selection.

It is still open to the materialist to retreat to an even more skeptical position. Yes, he might concede, we do actually have the conviction in some cases that we know something with absolute certainty, or know that something is true of necessity. But perhaps all such convictions are just illusions or feelings planted in us by nature. They are chemical moods, so to speak. For some reason our brains were fashioned by natural selection to have these feelings of certainty because they help us get through life. In Chesterton's words, they are just movements in the brain of a bewildered ape. This is a possible position, but it means, ultimately, abandoning all belief in human reason. I would rather take my stand with the mathematician G. H. Hardy, who said, "317 is a prime number, not because we think it so, or because our minds are shaped in one way rather than another, but *because it is so*, because mathematical reality is built that way."[21] And with Galileo, who said, "It is true that the divine intellect cognizes mathematical truths in infinitely greater plenitude than does our own (for it knows them all), but of the few that the human intellect may grasp, I believe that their cognition equals that of the divine intellect as regards objective certainty, since man attains the insight into their necessity, beyond which there can be no higher degree of certainty."[22]

I have cited the capacity to understand universals, or abstract concepts, openness to truth, the ability to attain certainty, and the power to recognize that some truths are true "of necessity," as being beyond the capacity of any merely material system, including the kind that the materialist conceives us to be—a computer programmed by natural selection. Another human intellectual ability involves all of these at once, namely the power to recognize that some truths hold in an infinite number of cases. Roger Penrose gives the following example.[23] We all know that $3 \times 5 = 5 \times 3$. As Penrose notes, mathematically speaking this is not the empty statement it appears to be. It really says that three groups of five objects and five groups of three objects contain the same number of objects. There is a simple pictorial argument that shows this: If we arrange fifteen objects in a three-by-five rectangular array, we see that it has five columns of three objects and three rows of five objects. Now, most of us, when we look at that array, will immediately "see" that the same thing works for a rectangular array of any size, so that $a \times b = b \times a$, in general, for any numbers a and b.

The ability to "see" this is a remarkable thing, as Penrose points out. We are seeing at once the truth of an infinite number of statements. Exactly the same observation was made by St. Augustine in his great philosophical work *De Libero Arbitrio*, completed around 395 A.D. In reference to a similar arithmetical statement he asked, "How do we discern that this fact, which holds for the whole number series, is unchangeable, fixed, and incorruptible? No one perceives all the numbers by any bodily sense, for they are innumerable. How do we know that this is true for all numbers? Through what fantasy or vision do we discern so confidently the firm truth of number throughout the whole innumerable series, unless by some inner light unknown to the bodily senses?"[24]

In this chapter I have used mostly mathematical examples. This is not essential to the arguments. Mortimer Adler's Aristotelian argument, discussed earlier, applies to any universal concepts, not just mathematical ones. Similarly, there are things we know with certainty that are not mathematical in nature. Nevertheless, mathematics is a particularly pure example of abstract thought. In the next chapter we shall look more closely at what is involved in it.

If Not the Brain, Then What and How?

A lot of people who might admit that materialism has difficulties explaining the human intellect nevertheless embrace it because they simply see no alternative. Let us grant, they say, that something we call the intellect exists and that it has the ability to understand abstract ideas. Where, they ask, does this intellect reside, if not in the brain? Is it floating around somewhere in space? And how, they ask, does the intellect work? There must be some mechanism by which it operates. But if it is by some mechanism, then why deny that that mechanism takes place in the brain and is a physical mechanism?

These seem like reasonable questions, but it is not clear that they will ultimately turn out to be meaningful. There are many things that we explain in terms of more elementary constituents or concepts. For example, the properties of a gas are explained by saying that the gas is made up of molecules. The pressure that the gas exerts on its container is the result of countless molecules bouncing off the container's sides. The temperature of the gas is a measure of how energetically its molecules are bouncing around. The sound that propagates in the gas is explained as waves of compression and rarefaction reducing or increasing the average spacing of the molecules. This kind of explanation of one thing in terms of something more basic is not always possible, however. It cannot be, since there must be some things that are the most basic of all and in terms of which everything else is explained. Until we have complete understanding, we will not know whether certain things we now treat as basic are reducible or not to something more basic still.

The first part of physics that was well understood was mechanics, which deals with the motion of material bodies. When electromagnetic phenomena began to be studied, therefore, it was natural to make "mechanical models" of electromagnetic fields. Only later was it realized that these "aether theories," on which a great deal of ingenuity was expended, were largely a waste of time. Electromagnetic fields were recognized to be a reality in their own right, as basic as material bodies. They were something qualitatively new, and the effort to find a material aether that explained them was seen to be misguided.

Of course, electromagnetic fields are just as physical as the material bodies studied by Newtonian mechanics, and we now have an overarching, coherent physical theory in which both electromagnetic fields and material bodies find their proper place. The point, however, is that attempts to find a coherent picture of reality by shoehorning new and poorly understood phenomena into existing conceptual categories, as the aether theorists did, are sometimes intellectual dead ends.

It is conceivable that all mental realities, including "intellect," "understanding," abstract "concepts," and so on, are ultimately explicable in terms of neurons firing, or some other basic physical events or entities, just as the phenomena of gases are explicable in terms of molecular motions. But it is just as conceivable that these phenomena involve something new, something that exists in its own right, and is not reducible to or explicable in terms of something more basic than itself. Science gives us examples of both possibilities.

But even if the intellect is not simply a material system, we would still want to know how it works. However, we must again be careful. It sometimes makes sense to ask "how" something works, in the sense of seeking a "mechanism" by which it happens, but sometimes it does not. For example, one can answer the question of how a phonograph produces sound, or how a television produces a picture. But it is not clear that it makes sense to ask "how" a mass produces a gravitational field, say. In Newton's theory, "mass" and "gravitational field" are fundamental concepts. Newton's law of gravitation posits the existence of a relationship between them and gives a quantitative account of that relationship. But it does not explain "how" the mass produces the field. As Newton himself said in the famous concluding words of his *Principia*: "I have not been able to discover the cause of those properties of gravity from phenomena, and I frame no hypotheses. . . . [It] is enough that gravity does really exist and act according to the laws which we have explained, and abundantly serves to account for all the motions of the celestial bodies, and of our sea."[25] Similarly, Einstein's theory, while it gives a deeper understanding of what a gravitational field is, namely the curving of space-time, does not explain "how" a massive body causes that curvature, in the sense of a mechanism.

All we know about the human intellect is that it is capable of having insights, of understanding meanings. By what mechanism? "How" does the intellect act on the physical brain? What is the intellect made of? Far from requiring a materialist answer, these questions may not even turn out to make any sense. Sometimes we must have the patience to hold certain questions in abeyance until we have the conceptual equipment and level of understanding that allows us to distinguish the good questions from the bad ones. To repeat again the wise words of Hermann Weyl: "One of the great differences between the scientist and the impatient philosopher is that the scientist bides his time. We must await the further development of science, perhaps for centuries, perhaps for thousands of years, before we can design a true and detailed picture of the interwoven texture of Matter, Life, and Soul."

22 | Is the Human Mind Just a Computer?

WHAT A COMPUTER DOES

As we saw in the last chapter, the materialist conceives of the human mind as being no more than a computer in operation. In this chapter I am going to explain some arguments, due to the philosopher John R. Lucas and the mathematician Roger Penrose, and based on a powerful theorem proved by the logician Kurt Gödel, that attempt to show that this conception cannot be right.

What is a computer? The name itself tells us: it is a device that computes. Computers come in two types, analog and digital. The difference is somewhat like the difference between an hourglass and a digital clock. The hourglass measures time by using a physical or mechanical process, while the digital clock reduces time to numbers. At one time, very sophisticated analog computers were built—for example, they were used to aim the huge guns on warships of World War II vintage. However, since the advent of electronics, almost all computers are digital. For the present discussions we can restrict our attention to digital computers, since the issue which concerns us has to do with what computers are capable of doing, and anything which could be done by an analog computer can be simulated by a digital computer.

Digital computers manipulate numbers. The numbers they display for us are usually in the decimal, or base-10, notation that we are taught in school. This is true of the displays of most pocket calculators, for instance. The numbers that computers use internally, however, are generally in binary, or base-2, form. That is, instead of using the digits 0 through 9, they use the two "bits" or "binary digits" 0 and 1. These are not actually written out as "0" and "1" inside the computer, of course; rather, 0 and 1 may be represented by some voltage having one value or another, or by some small piece of material being magnetized or not magnetized. This illustrates an important point: computers deal with information that is expressed in some kind of symbols, and it does not matter what particular symbols are used.

Rather than saying that computers manipulate numbers, then, it would be more accurate to say that they manipulate symbols. The symbols could stand for numbers, but they could also stand for words, or moves in a chess game, or even nothing at all. Computers manipulate these symbols using a rote procedure, called a "program" or "algorithm." At every step of its operations, the program or algorithm tells the computer exactly which manipulations to perform.

The idea of an algorithm may seem foreign to those who have not worked with computers, but actually almost everyone is familiar with certain simple algorithms. The procedures we all learned in grade school for doing addition, subtraction, multiplication, and long-division are algorithms. In doing a long-division problem, for example, we do a sequence of simple steps involving the symbols 0 through 9 in accordance with a well-defined recipe. The "inputs" are the numbers being divided, and the "output" is their quotient.

Now, for obvious reasons, mathematicians are very interested in this business of manipulating symbols using algorithms. Not only can mathematical calculations be done in this way, as we just saw, but so can mathematical proofs. In a proof one starts with certain statements that are assumed to be true—these are either unproven "axioms" or theorems that have been proven previously—and derives from them some new theorem using certain "rules of inference." A rule of inference is a logical or mathematical rule that enables one to deduce statements from other statements. All of this can be done by computers. First, the axioms and theorems have to be expressed in some symbolic language as strings of symbols that the computer can manipulate, and then the rules of inference have to be reduced to precise recipes for manipulating strings of symbols to produce new strings of symbols.

Let us give a simple illustration. Suppose that one of the axioms is that $x = x$, where x stands for any string of symbols. This axiom would tell us, for example, that $2 = 2$, and $17 = 17$, and $a + 2 = a + 2$ are true statements. And suppose that one of the rules of inference is that $x + y$ on one side of an = sign can be replaced by $y + x$, where x and y are strings of symbols.[1] For example, this rule of inference tells us that if we see $2 + 5$ in a true statement, we may replace it with $5 + 2$ and the statement will remain true. Similarly, we can replace $a + b$ with $b + a$. Now suppose that the theorem we are trying to prove is that $a + b + c = c + b + a$. One way to proceed (but not the shortest) is to take the following series of steps. Start with

$$a + b + c = a + b + c.$$

This is a true statement because of the axiom $x = x$, where x in the axiom is taken to be $a + b + c$. Next, on the right-hand side of the =, replace $b + c$ with $c + b$ (an application of the rule of inference):

$$a + b + c = a + \underline{b + c}.$$
(switch)

This gives

$$a + b + c = a + c + b.$$

Then, use the rule of inference again to replace $a + c$ on the right-hand side of the = with $c + a$:

$$a + b + c = \underline{a + c} + b.$$
(switch)

This gives

$$a + b + c = c + a + b.$$

Finally, replace $a + b$ with $b + a$ on the right:

$$a + b + c = c + \underline{a + b}.$$
(switch)

This gives the final result

$$a + b + c = c + b + a,$$

which is the result that we set out to prove.

This is a very simple example, but in fact even difficult proofs in mathematics can be done by such mechanical methods of symbol manipulation. What this indicates is that at least some forms of reasoning can be reduced to routine steps that can be carried out by computers. But does this mean that a computer can have the "power of reason" in the same sense that we do? Can a computer have an intellect?

In thinking about this question, it is important to keep in mind that there is a distinction between being able to manipulate symbols correctly according to some prearranged scheme and understanding the meanings of those symbols. A very good illustration of this is provided by the simple little proof that we just gave of the formula $a + b + c = c + b + a$. What does this formula mean? Well, an obvious meaning is suggested by the conventional usage of the symbols

+ and =. Interpreted in this way, the formula is a statement about adding numbers. However, the formula could just as well be interpreted in other ways. For example, the symbol + could be taken to stand for multiplication, in which case the axioms, the rules of inference, and the entire proof are just as valid, as the reader can easily check. In fact, the symbols in the formula might have nothing to do with arithmetic at all. They might stand for words of the English language. The symbol = might stand for any form of the verb *to be*; *a*, *b*, and *c* might stand for nouns; and + might stand for a conjunction like *and*. In that case, the axioms, rules of inference, and proof are also valid. The formula $a + b + c = c + b + a$ could then mean that "Tom and Dick and Harry are Harry and Dick and Tom," or "bell and book and candle are candle and book and bell."

We see, then, that reducing a process of reasoning to the level of mechanical symbolic manipulation has the effect of draining it of most if not all of its meaning. What is left is form without specific content. That is why such manipulations are called "formal." When mathematicians reduce a branch of mathematics to such manipulations they say they are "formalizing" it, and the resulting system, with its symbols, rules for stringing symbols together into formulas, axioms, and rules of inference, is called a "formal system." Computers operate on this formal level, which is one reason that they cannot have any understanding of the meanings of the symbols they manipulate. A computer does not know whether the symbols it is printing out refer to numbers or to Tom, Dick, and Harry. It is therefore only in a very restricted sense that we can say that computers "reason." What they can do is the mechanical parts of reasoning, which involve no understanding of meaning and therefore do not require intellect.

When we speak of "understanding" we put ourselves in much the same position with respect to the materialist as when we speak of free will. For even though materialists, as much as anyone else, actually do understand things, including abstract concepts, they profess not to know what words like *understand* mean. In order to be admitted into their lexicon such words must be defined in what the materialist regards as a suitably "scientific" way. But as we do not yet have a theory that explains how we understand abstract concepts, and we cannot fashion probes that can be inserted into brains or computers to detect the presence of abstract understanding, it is hard to satisfy the demand put on us to define "scientifically" what we mean. Of course, this demand is unreasonable, since all scientific statements ultimately rely for their meaning (as Niels Bohr emphasized) on other statements made in everyday language whose meaning is derived from ordinary experience. One could not teach science to a person using only equations or technical vocabulary. Nevertheless, in dealing with materialists we tend to reach an impasse, since in speaking about phenomena, however much a part of ordinary experience, which are not easily or at all reducible to numbers, we are at risk of being accused of speaking of unreal things.

To make any headway, then, we must start not with the question of what computers can "understand," but with what they can do. That is, what can a computer compute? What can it give as output? Are there certain problems to which a computer is incapable of printing out an answer, but to which a human being can supply one? This is the kind of question that is involved in the argument put forward by the philosopher John Lucas, and later refined by the mathematician and physicist Roger Penrose, that aims to prove that human mental processes are not simply those of a computer. This argument takes as its starting point a powerful result in logic called Gödel's Theorem.[2] Gödel's Theorem, as originally formulated, referred not to computers but to "formal systems," that is, to branches of mathematics that had been completely reduced to symbolic language. However, there is a very intimate connection between formal systems and computer programs or algorithms. Both are essentially mechanical schemes of symbol manipulation. In fact, several years after Gödel proved his famous theorem about formal systems, similar theorems were proven about computers by Alan Turing and others. I shall therefore not bother to distinguish in the following discussions between formal systems and computer programs.

It is not hard to explain the gist of the Lucas-Penrose argument, as I already noted in chapter 3. In essence, Gödel showed that if one knew the program that a computer uses, one can in a certain sense outwit the computer. What Lucas and Penrose argued is that if we ourselves were just computers we would be able to know our own programs and thus *outwit ourselves*, which is clearly not possible. This is the Lucas-Penrose argument in a nutshell. In what follows, I shall give a more precise account of both of Gödel's Theorem and the Lucas-Penrose argument. In appendix C, I give a more mathematical account of how Gödel proved his theorem.

WHAT GÖDEL SHOWED

The theorem proved in 1931 by the Austrian logician Kurt Gödel is about formal systems that have at least the rules of arithmetic and simple logic built into them. Such systems can be either "consistent" or "inconsistent." This is a very important distinction to understand, because Gödel's Theorem applies only to consistent formal systems. A system is called inconsistent if it is possible, using its rules, both to prove some proposition and to prove its contrary. For example, arithmetic would be inconsistent if we could prove both that $a = b$ and that $a \neq b$. (The symbol $\neq$ means "is not equal to.") A system is called consistent, on the other hand, if no such contradictions can arise in it.

There is a very important fact about inconsistent formal systems: if *any* contradictory statements can be proven, then *all* contradictory statements can be proven. This may be surprising, but it actually follows from the rules of elementary logic. Rather than explaining how one shows this in general, I will give an example. Suppose I set up the rules of arithmetic (badly) so that I can prove that $1 = 0$. Using that one inconsistency, I can prove all arithmetic statements, whether they are true or false. For example, I can prove that $13 = 7$. To do that, I just take $1 = 0$, and multiply both sides by 6, to get $6 = 0$. Then I just add 7 to both sides to get $13 = 7$.

In other words, a formal system cannot be just a little inconsistent; if it is inconsistent, it is inconsistent all the way, thoroughly, radically, completely. In an inconsistent formal system *anything* that can be stated in that system can be proven using its rules.

There is an amusing story which illustrates this logical fact and which also shows how quick-witted the philosopher Bertrand Russell was. Russell, who did important work in logic as well as philosophy, was asked by an interviewer whether, given the fact that anything can be proven in inconsistent systems, he could use the statement "$2 = 1$" to prove that he (Russell) was the pope. Russell instantly proceeded to do just that. Consider (he said) a room containing 2 people, namely Bertrand Russell and the pope. Now, since $2 = 1$, it is also true to say that there is only 1 person in the room. Since Russell is in the room, and the pope is in the room, and there is just 1 person in the room, then Russell and the pope must be the same person.

From the fact that *any* proposition which can be stated in an inconsistent formal system can be proven using its rules, a rather surprising conclusion follows: If we can find even one proposition that can be stated in a formal system but that *cannot* be proven using its rules, then we know that that system is completely consistent.

Having this basic distinction between consistent and inconsistent formal systems under our belts, we can say what it is that Gödel proved. What Gödel showed[3] is that in any consistent formal mathematical system in which one can do at least arithmetic and simple logic there are arithmetical statements which can neither be proved nor disproved *using the rules of that system* (i.e., using its axioms and rules of inference), but which nevertheless are in fact true statements. Statements that can neither be proved nor disproved using the rules of a formal system are called "formally undecidable propositions" of that system.

However, Gödel did much more than this. He also showed, for any particular consistent formal system containing logic and arithmetic, how to actually find one of its formally undecidable but true-in-fact propositions. The particular one he showed how to find is called the "Gödel proposition."

To sum up, if **F** is any consistent formal system that includes logic and arithmetic, then Gödel showed how to find a statement in arithmetic, which we may call G(**F**), that is *neither provable nor disprovable using the rules of* **F**; and he further showed that G(**F**) *is nevertheless a true arithmetical statement.*

What Gödel proved can be carried over to apply to computer programs. For a computer program **P** that is known to be consistent, and that is powerful enough to do arithmetic and simple logic, one can find a statement in arithmetic, G(**P**), that cannot be proven or disproven by that program. And one can show that G(**P**) is a true statement. It is in this sense that one has "outwitted" the computer, for one has succeeded in showing that a certain proposition is true that the computer itself cannot prove using its program.

Gödel proved one more thing in his famous theorem.[4] He showed that the consistency of a formal system is itself undecidable using the rules of that system. That is, if a formal system (or computer) is consistent, it cannot prove that it is.

THE ARGUMENTS OF LUCAS AND PENROSE

In 1961, John R. Lucas, a philosopher at Oxford University, set forth an argument, based on Gödel's Theorem, to the effect that the human mind cannot be a computer program.[5] He wrote,

> Gödel's theorem seems to me to prove that Mechanism is false, that is, that minds cannot be explained as machines. So also has it seemed to many other people: almost every mathematical logician I have put the matter to has confessed to similar thoughts, but has felt reluctant to commit himself definitely until he could see the whole argument set out, with all objections fully stated and properly met. This I attempt to do.[6]

We shall explain the Lucas argument in a moment. First let me say a few words about its history. Gödel himself, though he did not lay out the details of an argument like Lucas's in public, seems himself to have reached the same conclusion. He did not believe that the human mind could be explained entirely in material terms. In fact, he called that idea "a prejudice of our times."[7] In this he was only too right; the prejudice that the mind is a computer has hardened in many minds to become a dogma. It was only to be expected, therefore, that Lucas's argument would be generally rejected when he advanced it in the 1960s, and that it would have little impact on the thinking of people who work in the field of artificial intelligence.

Recently, however, the eminent mathematician and mathematical physicist Roger Penrose (who also happens to be at Oxford) has revived Lucas's argument. His first book on the subject, *The Emperor's New Mind*, published in 1989,[8] provoked even more criticism than Lucas had. Much of this appeared in the journal *Behavioral and Brain Sciences* in 1990,[9] and Penrose answered it in a second book, *Shadows of the Mind*,[10] which also stimulated much debate.[11] One can best sum up the situation by saying that, while no one has succeeded in refuting the Lucas-Penrose argument, it has not succeeded in changing many minds. As we shall see, there are escape clauses in the argument that the materialist can make use of, but at the expense of diminishing the plausibility of his entire point of view. (It should be noted that Penrose, unlike Gödel and Lucas, seems to be a materialist himself, though an unusually open-minded one. He argues that while the human mind is not simply a computer it might still be explicable in physical terms in some as yet undreamt-of way.)[12]

Now I will explain the Lucas argument. First, imagine that someone shows me a computer program, **P**, that has built into it the ability to do simple arithmetic and logic. And imagine that I know this program to be consistent in its operations, and that I know all the rules by which it operates. Then, as proven by Gödel, I can find a statement in arithmetic that the program **P** cannot prove (or disprove) but which I, following Gödel's reasoning, can show to be a true statement of arithmetic. Call this statement G(**P**). This means that I have done something that that computer program cannot do. I can show that G(**P**) is a true statement, whereas the program **P** cannot do so using the rules built into it.

Now, so far, this is no big deal. A programmer could easily add a few things to the program—more axioms or more rules of inference—so that in its modified form it can prove G(**P**). (The easiest thing to do would be simply to add G(**P**) itself to the program as a new axiom.) Let us call the new and improved program **P'**. Now **P'** is able to prove the statement G(**P**), just as I can.

At this point, however, we are dealing with a new and different program, **P'**, and not the old **P**. Consequently, assuming I know that **P'** is still a consistent program, I can find a Gödel proposition for *it*. That is, I can find a statement, which we may call G(**P'**), that the program **P'** can neither prove nor disprove, but which I can show to be a true statement of arithmetic. So, I am again ahead of the game.

However, again, this is no big deal. The programmer could add something to **P'** so that it too could prove the statement G(**P'**). By doing so he creates a newer and more improved program, which we may call **P''**. This race could be continued forever. I can keep "outwitting" the programs, but the programmer can just keep improving the programs. Neither I nor the programs will ever win. So, we have not proven anything. But here is where Lucas takes his brilliant step. Suppose, he says, that I myself *am* a computer program. That is, suppose that when I prove things there is just some computer program being run in my brain. Call that program **H**, for "human." And now suppose that I am shown *that* program. That is, suppose that I somehow learn in complete detail how **H**, the program that is *me*, is put together. Then, assuming that I know **H** to be a consistent program, I can construct a statement in arithmetic, call it G(**H**), that cannot be proven or disproven by **H**, but that I, using Gödel's reasoning, *can* show to be true.

But this means that we have been led to a blatant contradiction. It is impossible for **H** to be unable to prove a result that I am able to prove, because **H** is, by assumption, *me*. I cannot "outwit" myself in the sense of being able to prove something that I cannot prove. If we have been led to a contradiction, then somewhere we made a false assumption. So let us look at the assumptions. There were four: (*a*) I am a computer program, insofar as my doing of mathematics is concerned; (*b*) I know that program is consistent; (*c*) I can learn the structure of that program in complete detail; and (*d*) I have the ability to go through the steps of constructing the Gödel proposition of that program. If we can show that assumptions *b*, *c*, and *d* are valid, then we will have shown that *a* must be false. That is, we will have shown that I am not merely a computer program. This is Lucas's argument. Of course, there is nothing special about me. The same argument could be made about you or other human beings. But, in any event, as long as it applies to any human being, then that human being, at least, is not a mere computer.

From the foregoing, we see that Lucas's argument has possible loopholes, or avenues of escape, that can be used by those pitiable people who cling to the belief that they are computers. They can, instead of denying *a*, deny the assumptions *b*, *c*, or *d* of the proof. I have given here the argument as Lucas originally gave it. Roger Penrose presents it in a version that is slightly different and in his view somewhat stronger. He has also given answers to the numerous objections that have been raised against it.

AVENUES OF ESCAPE

I cannot possibly discuss here every version of every objection that has been raised to the Lucas-Penrose argument or every way of attempting to escape its conclusions. I will, however, try to make clear what the main issues are.

Let us start with the avenue of escape for assumption *c*: that a human being cannot know his own computer program. As a practical matter, this is undoubtedly true. In the first place, to understand the structure of any particular human brain in complete detail would almost certainly entail procedures that would be so invasive as to destroy the brain in question. Thus, it is hard to see how a person could know his *own* brain's program. Nevertheless, a person might be able to discover the program of someone else's brain, and it might be argued that all human brains work in basically the same way; but this last point can certainly be disputed.

However, I think there is a much better answer. If our thinking is really just the running of a computer program, it does not matter for the Lucas argument what kind of machine that program is run on. The same program can run on a computer that uses vacuum tubes, or silicon chips, or neurons. One can certainly imagine a world in which the very same program that my brain uses is run on a kind of machine whose internal parts and programming are easily open to self-inspection. That machine (according to the materialist viewpoint) ought to be able to prove the same things that I can prove. But that machine, unlike myself, could know its own structure without any invasive or destructive procedures. The Lucas-Penrose argument can be applied to that machine to reach the same contradiction as before, but without escape avenue *c*.

Now let us turn to the escape avenue for assumption *d*. Suppose that I am able to have access to complete information on the structure of my brain's program. Is it not possible that the quantity and the complexity of that information is so great that I would die of old age before I was able to analyze it and construct the appropriate Gödel proposition for it? Again, this objection is based upon the details of how my body and brain are constructed, and in particular upon the fact that I will die of old age. There is no reason, however, that my program could not, in principle, be run on a machine that was far more durable and reliable in an engineering sense than I am.

A more significant objection is based on the possibility that the Gödel proposition of the human program, $G(H)$, is so complex that the human mind simply does not have the capacity to construct it. This is another version of escape avenue *d*. This objection is based on a possible limitation of the human program itself rather than of the human machine that runs it. The idea that the finiteness of the human mind might prevent it from analyzing its own program in the way supposed in the Lucas-Penrose argument has some plausibility. The issue has to do with how complicated $G(H)$ is. This is a quantitative question, and can therefore be analyzed mathematically. Penrose has done this and concludes that the degree of complexity of $G(H)$ would be virtually the same as that of H itself, so that if a brain can learn what its own program is, it can probably also do the computations necessary to construct its Gödel proposition.[13]

Aside from the technicalities of Penrose's analysis of this question, there is a more basic consideration. How likely is it that natural selection could have produced a program so complicated that it defies human analysis even given an unlimited amount of patience and time? (Recall that we can always do the Lucas-Penrose argument assuming that the human program is run on a machine that does not wear out.) Our forebears are supposed to have diverged from those of chimpanzees about 6 to 8 million years ago. So within a few hundred thousand generations humans developed the ability to do abstract mathematical reasoning. In fact, that breakthrough certainly happened much more recently. The genus *Homo* has only been around for about 2 million years, and *Homo sapiens* has only been around for a few hundred thousand years, or about ten thousand generations.) Moreover, from generation to generation the changes in the structure of the brain and its program were presumably fairly slight, and these changes were governed by a haphazard process of trial and error. It seems highly implausible, therefore, that a physical structure constructed in this blundering way in a limited period of time could not, even in principle, yield to an intelligent analysis which had unlimited time at its disposal. (Again, we are assuming access to complete information about the actual physical structure of the brain.)

We come now to the escape avenue for assumption *b*. The most popular objection to the Lucas-Penrose argument seems to be that human beings are inconsistent computer programs, or at least do not know themselves to be consistent. At first glance this seems incontestable. After all, who is it that has never reasoned inconsistently or made mistakes in mathematics?

In response to this objection two points must be made. The first is that there is a difference between a computer that is using a truly inconsistent program and one that, though using a consistent program, malfunctions because of interference with its normal scheme of operation. A computer which has a perfectly consistent program for doing arithmetic may nevertheless give a wrong answer if, for example, a cosmic ray switches a bit from "0" to "1" in its memory, or if some circuit element or chip develops a physical flaw, due perhaps to excessive heat. It seems plausible to suppose that many human errors, such as those that arise from fatigue or inattention, are due to malfunctioning.

The second point is that if we are programs that are really, in themselves, inconsistent then we must be radically inconsistent, as I explained before. A truly inconsistent program or formal system can prove *anything whatever*. Indeed, it is just for that reason that it evades the Lucas-Penrose argument. That argument was based upon the inability of **H** to prove G(**H**); but if **H** is inconsistent it can prove *anything*. Thus, an inconsistent program is vastly more powerful than a consistent one, just as a liar is able to assert more things than an honest person can. But while more powerful, in a sense, an inconsistent program is also quite helpless, for *it cannot recognize its own mistakes*. An inconsistent program would be just as able to prove that $2 + 2 = 17$ as that $2 + 2 = 4$, and would not have any way to tell which answer to prefer. It would, therefore, be equally satisfied with either result. It is quite otherwise with humans. We do make mistakes in our sums, but we can also spot those mistakes, or have them pointed out to us, and recognize that they *are* mistakes. As Lucas observed,

> If we are really inconsistent machines, we should remain content with our inconsistencies, and happily affirm both halves of a contradiction. Moreover, we would be prepared to say absolutely anything—which we are not. . . . This surely is a characteristic of the mental operations of human beings: they are selective; they do discriminate between favoured—true—and unfavoured—false—statements; when a person is prepared to say anything, and is prepared to contradict himself without any qualm or repugnance, then he is adjudged to have "lost his mind." Human beings, although not perfectly consistent, are

not so much inconsistent as fallible. A fallible but self-correcting machine would still be subject to Gödel's results. Only a fundamentally inconsistent machine would escape.[14]

Again, it must be emphasized that an inconsistent system has no way to resolve contradictions. In such a system $2 + 2 = 4$ is in no way better than $2 + 2 = 17$ or any other result. It is possible that an inconsistent machine could be prevented from ever actually making explicitly contradictory statements: it could have a "stop order," so that if it ever said or printed out "$2 + 2 = 5$" it would prevent itself from later saying or printing out "$2 + 2 \neq 5$." Such a machine could never be "caught" in a contradiction, but would still be radically inconsistent in a way that humans are not. Again to quote Lucas:

> No credit accrues to a man who, clever enough to see a few moves of argument ahead, avoids being brought to acknowledge his own inconsistency, by stonewalling as soon as he sees where the argument will end. Rather, we account him inconsistent too, not, in his case, because he affirmed and denied the same proposition, but because he used and refused to use the same rule of inference. A stop-rule on actually enunciating an inconsistency is not enough to save an inconsistent machine from being called inconsistent."[15]

Humans, therefore, have an ability to reason consistently in a way that an inconsistent machine would not have and would not be able to fake.

There is another possibility, and that is that the human program, **H**, has built into it a rule which is inconsistent, but which only comes into play in some very esoteric situations which rarely come up, or perhaps have never yet come up in human experience. A mathematician, for example, might be happily going along thinking that $1 \neq 0$, when suddenly he would be confronted with an argument that called into play that inconsistent rule that was lurking in his brain's program, and this argument would convince him that $1 = 0$ after all. Nor would he be able to find his mistake, for by the rules followed by his mental processes — the rules that define what seems "rational" to him — there would be no mistake. The argument that proved that $1 = 0$ would always seem quite reasonable to him. Indeed, because he is inconsistent, *any* proposition would be provable to him by some argument that he would find perfectly valid.

If human beings were like this, then we would obviously be utterly irrational in the final analysis. All human reasoning would be an exercise in futility. Here we get down to the basic issue: Are we rational, and do we know that we are?

It is not a question whether human beings are sometimes irrational. As Lucas observed, there is a difference between fallibility and radical inconsistency. The question is whether we can *ever* be truly rational and *know* that we are being rational. Is there ever a time when I can be sure that a particular statement is true or piece of reasoning is valid and that the contrary is not? I assert that there is. For instance, I claim to know that 1 is not equal to 0. Moreover, I also claim to know that no argument that I would recognize as a valid one would ever lead me to conclude that 1 is equal to 0. Or, rather, if I ever did fall for some fallacy that implied that $1 = 0$, I would at least be able to recognize it as a fallacy. This I claim to know about myself, and I believe that any reasonable person knows this about himself too. And if I know this, then I know that I am not an inconsistent program.

What, then, do I say to someone who points out that I have often claimed to know something in the past only to have it shown that I was in error? If my certainty was illusory in the past, then is it not possible that all my certainties may be illusory? Not necessarily. We often speak rather loosely about certainty. In many cases we mean simply a practical or "moral" certainty, as in our certainty that the Sun will come up tomorrow, not a real certainty as in a certainty that $1 \neq 0$. In addition, there are times when we are delusional. When asleep, for example, we sometimes dream that we are awake. In fact, in our dream we may even feel quite confident that we are awake. Notwithstanding this, I think few people would deny that there are times when we really do know that we are awake, and know that we are not deluded in thinking so. This is a paradoxical fact, perhaps, but a fact nonetheless.

One of the amusing aspects of the debate about the Lucas-Penrose argument is that many of those who claim that human beings are inconsistent programs, and who therefore are committed to the view that genuine certainty is impossible, nevertheless act as though they are quite certain of one thing—namely that the Lucas-Penrose argument is wrong!

It is very interesting that one of the most common objections made against the Lucas-Penrose argument involves the claim that human beings are fundamentally inconsistent. It is strange that in making this objection many materialists feel that they are fighting the good fight against what they regard as the superstitious idea of a "soul." It used to be that those who rejected religious tenets usually did so in the name of human reason. They called themselves "rationalists." But the new kind of skepticism is willing, in order to debunk the spiritual in man, to call into question human reason itself. According to this view, we are not even supposed to be able to trust ourselves about the simplest truths of arithmetic. G. K. Chesterton, with prophetic insight, saw where things were heading almost a century ago:

> Huxley preached a humility that is content to learn from Nature. But the new sceptic is so humble that he doubts if he can even learn. . . . We are on the road to producing a race of men too mentally modest to believe in the multiplication table. We are in danger of seeing philosophers who doubt the law of gravity as being a mere fancy of their own. Scoffers of old were too proud to be convinced; but these are too humble to be convinced.[16]

23 | What Does the Human Mind Have That Computers Lack?

The Lucas-Penrose argument is an attempt to establish that the human mind is not merely a computer. But does it tell us what, if anything, the human mind possesses that computers lack? It at least suggests an answer.

Let us recall one of the things that Gödel proved. He showed that the Gödel proposition for a formal system is a true one even though it is "formally undecidable." At first glance that may seem contradictory: If a proposition is formally undecidable, then how can Gödel or anyone else decide that it is true? The answer lies in what is meant by "formally undecidable." To say that a proposition is formally undecidable in a certain system means only that it cannot be proven or disproven *using the rules of that system.* It may, nevertheless, be provable using some form of reasoning that has not been written into the rules of that system.

Let us take a simple example. Suppose that the only thing that a certain computer program can do is add and subtract numbers. Such a program would be able to prove that the number 20 can be gotten by adding two odd numbers. It could do this very easily by, for instance, adding 7 and 13 and getting 20. However, the program would be incapable of settling the issue of whether 20 can be gotten by adding *three* odd numbers together. It could keep trying various triplets of numbers and show that in each case they do not add up to 20, but obviously that would not settle the issue, since there would always be an infinite number of other triplets to try. (I am including both positive and negative numbers. Otherwise, there would be only a finite number of cases for the adding machine to try.)

That 20 cannot be obtained by adding three odd numbers is true, but it is undecidable by a program that can only add and subtract. How do *we* know that it is true? We know because we are able, not only to add and subtract, but to *reason about* the process of adding and subtracting numbers. The program follows certain rules, but it cannot look at those rules from outside, so to speak, and understand what it is about the structure of those rules that makes it impossible to get 20 from three odd numbers.

Perhaps the point can be made clearer by an analogy. Imagine someone wandering inside a complicated maze, and unable to find the exit. It may be, in fact, because of the way the maze is designed, that the exit cannot be reached from where the person is inside it. However, from inside the maze, this fact may not be apparent. (It may even be impossible to prove by wandering about inside the maze, if the maze is infinitely large.) However, to someone who can look down on the maze from above, and see where the exit is, where the person is, and how the maze is constructed, it might be obvious that the person in the maze cannot reach the exit from where he is no matter what he does. The observer from above might see, for example, that the part of the maze that has the exit is walled off from the part of the maze where the man is wandering, as in figure 14.

Figure 14.   A man in an infinite maze may be unable to reach the exit because it is walled off completely from him. This may be obvious to an observer looking down on the maze from the outside, but impossible to prove by exploring the maze from within.

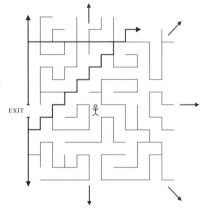

The analogy is this: The man in the maze is the computer. The paths of the maze are the rules of the computer program, which guide the steps of the computer. Just as the paths of the maze take the man from one place to another, so the rules of the program take the computer from one statement or string of symbols to another. The statement "the man cannot reach the exit from where he is by following the maze," is analogous to the Gödel proposition of the computer program. This statement may not be provable by the man (or disprovable by him) no matter how much he wanders in the maze. It is "undecidable" by merely wandering in the maze. But it is nevertheless true, and can be seen to be true by someone who understands how the maze is put together.

There are certain things that a consistent program cannot prove, because it is trapped within its own rules, as the man is trapped in the maze. But a mind that is not so trapped can examine the program *from outside*, as it were. And in that way it can gain insights that enable it to reach conclusions unavailable to the program itself. In fact, that is precisely how Gödel proved his theorem. He went outside the "formal system" and used methods of reasoning that would not be available to a mind that was forced to follow the rules of the formal system. And that is how he was able to show that the Gödel proposition is a true one, even though it is undecidable within the rules of the formal system.

What lies behind Gödel's Theorem, then, and the argument that Lucas and Penrose base on it, is the ability of the mind to go beyond a given specified set of rules and procedures to an insight into those rules. It is precisely the fact that we do not merely operate on the "formal" level, but also on the level of meaning and understanding. The machine can only "reason" to the extent that it can blindly follow some steps of reasoning that are on its list of allowable procedures. However, given any predetermined list of allowable steps of reasoning, Gödel showed that it is possible to go beyond them and reach conclusions that are not attainable by merely following those steps.

In other words, a formal system has certain procedures, including certain ways of reasoning, built into it. But Gödel showed that there are always *other* valid forms of reasoning that are *not* built into it. One can design more and more powerful formal systems (or computer programs) by building into them more and more ways of getting from one true statement to another, but Gödel showed that there would always be some left out. There are always steps of reasoning whose validity would appear "obvious" to a rational being, but that are not on the formal system's (or program's) list of steps that it can carry out. Penrose expresses the point in this way:

> One might imagine that it would be possible to list all possible 'obvious' steps of reasoning once and for all, so that from then on everything could be reduced to computation—i.e., the mere mechanical manipulation of these obvious steps. What Gödel's argument shows is that this is not possible. There is no way of eliminating the need for *new* 'obvious' understandings. Thus, mathematical understanding cannot be reduced to blind computation.[1]

The essential point, then, would appear to have to do with the power to "understand." In chapter 21, I argued that humans have this power of understanding while computers lack it. Certainly, most people would admit that this is true of the computers that exist today. However, there are many computer enthusiasts who would argue that the reason that present-day computers cannot "understand" is that their programs have not yet attained the necessary complexity and power. They point out that the human brain has about 100 billion neurons, many of which connect up to over a thousand other neurons. This makes the human brain far more complex than any computing device so far constructed artificially. It is this, many materialists believe, that is responsible both for the phenomenon of consciousness, which computers also lack (in most people's view), and for the human ability to understand. The arguments of Lucas and Penrose cast considerable doubt, and in the view of some disprove, the idea that mere numbers of neurons is the issue.

CAN ONE HAVE A SIMPLE IDEA?

Apart from the difficulties that Lucas's and Penrose's arguments create for it, the notion that the power of understanding is a matter of vastly complicated circuitry is implausible on its face. Why should a thousand neurons, say, be too little to give rise to the power of understanding abstract concepts, but a few billion be enough? It is certainly true that some of our mental operations require enormous computing power, and therefore a huge number of neurons. That is the case, for example, of vision.

As I write this, I am sitting on my back porch and am confronted by a very complex scene: trees with all their leaves, branches, and twigs; the lawn with all its blades of grass, dandelions, and clover; a swing set, with all its attachments and parts; two rabbit hutches with their occupants; clotheslines; neighbors' houses; patches of sky among the foliage; and so forth. In the vast number of details, there is an enormous amount of information that my visual system is capable of processing very quickly. Moreover, as I survey the scene, I do not see it merely as a mass of details. I see the details in their relationships to each other. I see the complex pattern of shapes and colors of the swing set *as* a swing set. I see the trees *as* trees. Thus the human visual system does a lot of interpretation. In fact, I cannot see without also interpreting to some extent what I see. Now all of this, which humans do much better than machines are so far able to do, clearly does require tremendously complicated circuitry. But that is because seeing is a complex thing.

However, it is not at all clear that acts of abstract understanding are necessarily complex in themselves. Consider what happens in understanding the statement "2 + 2 = 4," for example. What is involved in that mental act? Well, one must understand the number 2, the number 4, the notion of equality, and the notion of addition. This act of understanding would seem, therefore, to involve about four subsidiary acts of understanding. Why should that act, then, require billions of neurons? Indeed, how about the act of understanding the single concept "2"? At least naively, this does not appear to be something that is all that complex. In fact, it appears to be rather irreducibly simple. I can talk about seeing 95 percent of my backyard, or 90 percent, or 57 percent, or 2 percent. All I have to do is block out some of the view with my hand. If I do, there is less data to process and interpret. But what does it mean to have 57 percent of the concept "2"? Some acts of understanding seem to be quite simple, and it is far from clear why a vast apparatus with billions of parts should be required to produce them.

I do not at all wish to deny the fact that an enormous amount of neural activity occurs when we are engaged in understanding things. Our acts of understanding, even of abstract concepts, are usually accompanied by visualizations that assist our thinking. We all know that when we are straining to understand, many things are "called to mind" from memory, many things are analyzed, juxtaposed, compared, and sorted out. Obviously, the brain's circuitry must be very busy in all this. What is questionable, however, is the idea that the act of understanding itself—that flash of insight in which our strenuous attempts to understand often culminate—is a vastly complex phenomenon.

The essence of an act of understanding is to grasp the simplicity of something. Admittedly, what we are grasping may be a relationship among a large number of things, but the relationship is one thing and not a large number of things. For example, suppose we teach a child about numbers. We count: "one, two, three," and so on. There comes a time when the child gets the point and sees that there is an unending sequence of numbers. At that moment the child has grasped the concept of number, a single concept that embraces—comprehends—that whole infinite sequence of numbers.[2] The very word *comprehends* means literally "holds together," for an act of understanding holds a multiplicity of things together in a single insight. The expression "getting the point" is also very suggestive, for a point is something indivisible. The great American philosopher and psychologist William James observed that "no matter how complex the object may be, the thought of it is one undivided state of consciousness."[3]

It is an interesting fact that one cannot have two distinct acts of understanding at the same time. My mind can do many things at the same time: it can hear, see, smell, taste, feel, have a variety of emotions, and understand all at the same time. But it cannot have two unrelated insights at the same time. I can, indeed, have an insight that involves several other subsidiary insights, but only by grasping some unity that embraces them all, by having some overarching insight that brings them all together. When we learn a subject, our insights about it do not merely pile up, they somehow cohere to form a unified understanding. In other words, insight tends toward simplicity and unity, and an act of insight, like the "point" which is gotten, can be thought of as something indivisible.

Here I am talking about human acts of insight, but these reflections also shed some light on certain traditional ideas about the nature of God. This should not be too surprising, as our intellects are supposed to be made in the "image of God." God has been described by theologians[4] as the infinite act of understanding that grasps all of reality. (See appendix A, pp. 262–63.) Now, if we appreciate that even a finite human act of understanding or insight tends toward simplicity and is in some sense an indivisible unity, it helps us to understand what the mediaeval Jewish and Christian theologians, such as Maimonides and St. Thomas Aquinas, meant when they spoke of the perfect unity and "simplicity" of God.[5] Actually, this doctrine that God is utterly simple in his nature goes back to very ancient times. St. Irenaeus, for example, wrote in the second century:

> Far removed is the Father of all from those things which operate among men, the affections and the passions. He is simple, not composed of parts, without structure, altogether like and equal to himself alone. He is all mind, all spirit, all thought, all intelligence, all reason. . . . This is the manner in which the religious and pious are accustomed to speak of God.[6]

Is the Materialist View of the Mind Scientific?

Science begins with phenomena and tries to understand them. But it appears that many materialists, in thinking about mind, do quite the reverse. They start with a theory and dismiss certain facts.

The conviction, so widespread nowadays, that the human mind is nothing but a machine has its basis, I believe, in the following argument. Suppose there were something about the human intellect and will that was not reducible to matter alone. How could that something have come into existence? Before life evolved, there was just inanimate matter on Earth—on this the materialist and the religious believer agree. And before a child is conceived, the sperm and egg are just made up of matter. On this too there is agreement. How, then, could a "spiritual soul" make its appearance? If it cannot be reduced to the "merely material," then obviously it cannot be generated by merely physical processes such as evolution or sexual reproduction.

The religious believer accepts this logic, and attributes what is "immaterial" in the human being—in each human being—to a special creative act of God. While the physical body evolved and reproduces, the spiritual in a human being, i.e., that which goes beyond the physical, must be the result of a special act of creation. This is what is symbolically represented in Genesis by God "breathing" upon Adam. The atheist, however, turns this around: since there is no God (he says), there is no way that a human being can have come to be anything other than or in addition to a material system.

Thus the idea that man can be nothing other than a machine is really nothing other than a pure deduction from atheism. There is not a shred of positive evidence that a material system can reproduce the human abilities to understand abstractly and will freely. And we have seen, on the contrary, quite a few reasons to doubt that it is possible. However, having started with the empirically quite-unsupported postulate of atheism, the materialist is practically forced to call a variety of empirical facts "illusions"—not facts that are in front of his eyes, but facts that are behind his eyes, so to speak, facts about his own mind.

None of this is to deny that there are some very hard questions that arise from the idea that the human mind is not entirely reducible to matter. There certainly are. For instance, if there is something immaterial about the mind, how does it affect the brain and body? How is it integrated with the material brain to form one coherent and intelligible entity? In theological terms, how do "spirit and matter" in man form "not two natures united, but . . . a single nature"?[7] These are perplexing questions. But the price of following facts is often perplexity. It is not the way of science to avoid perplexities by denying facts. Rather, as Weyl observed, the way of the true scientist, as opposed to the "impatient philosopher," is to "bide his time."

24 | Quantum Theory and the Mind

In the previous chapters we discussed the question of whether the human mind is merely mechanical in the sense of being a mere automaton that follows fixed rules. Now we turn to the question of whether it is merely mechanical in the sense of being merely physical. Is the human mind nothing but matter in motion? Surprisingly, physics itself, a subject concerned only with matter and its properties, may tell us that the answer is no. At least, this is the view of some modern physicists of great eminence. Their view is based on one of the epoch-making discoveries of twentieth-century science: quantum theory. What they claim is that the fundamental postulates of quantum theory imply that the human mind cannot be completely reduced to physics.

Quantum theory was and is the most profound discovery about nature since Isaac Newton. It is not an isolated theory applying to this or that phenomenon. In fact, it is not so much a theory *in* physics, as an entirely new theory *of* physics. Developed from 1900 to 1925, it was a radical revolution in the foundations of the field. As might be expected of so fundamental a change, it was not the work of a single person but of many great scientists. The list of those who contributed key insights is awe-inspiring: Max Planck, Louis de Broglie, Albert Einstein, Niels Bohr, Erwin Schrödinger, Werner Heisenberg, Max Born, Wolfgang Pauli, Paul Dirac, John von Neumann, and a host of lesser, but still brilliant, lights.

Quantum theory is perhaps the most successful theory in the history of science. Over the course of seventy-five years it has been applied to a vast range of phenomena, and wherever it has been applied it has been found to work beautifully. In many cases its predictions have been tested to astonishing accuracy.[1] And yet, there is something disturbing about it, which caused even some of those who helped to develop it to be deeply dissatisfied. Einstein himself never came to terms with it. He wrote, "the more success quantum theory has, the sillier it looks." Schrödinger remarked in 1926, while visiting Bohr and Heisenberg, "If all this damned quantum jumping were really to stay, I should be sorry that I ever got involved with quantum theory."[2] (The quantum jumping has stayed.)

The problem with quantum theory lies neither in its mathematical formalism, which is simple and elegant, nor in its match to experiment, which is perfect, but in its philosophical implications. No matter how one looks at it, some of the things it says about the physical world seem quite bizarre. This has led to a variety of opinions as to how some of its basic concepts should be interpreted. I will be basing my discussion on the traditional interpretation, sometimes also called the "standard," "orthodox," or "Copenhagen" interpretation. Later I will discuss alternative views, such as the "many-worlds interpretation." I will treat them only briefly, however, as in my view they are in one way or another unsatisfactory.

There are two aspects of quantum theory that make it relevant to the nature of the human mind. The first is that quantum theory is inherently probabilistic in a radical way in which "classical physics" (i.e., pre-quantum physics) was not. As we saw in chapter 21, this is important for the question of whether physics leaves an opening for free will. The second aspect, which is closely related to the first, is that the so-called "observer" plays a key role in quantum theory—at least as traditionally interpreted. This role seems to place the observer on a level quite distinct from that of a mere material system. In particular, we shall see that certain logical problems arise if one tries to include the observer himself completely in the mathematical description of events. Since human beings are observers (and perhaps the only observers), this fact clearly has potentially huge implications for the question of whether the human mind is entirely reducible to physics and mathematics.

It was on the basis of the traditional interpretation of quantum theory that the late Sir Rudolf Peierls, one of the leading theoretical physicists of his generation, asserted that

> the premise that you can describe in terms of physics the whole function of a human being . . . including its knowledge, and its consciousness, is untenable. There is still something missing.[3]

Eugene Wigner, a Nobel Prize-winning physicist, in a classic essay on the implications of quantum theory, wrote that quantum theory is incompatible with the idea that everything, including the mind, is made up solely of matter:

> [While a number of philosophical ideas] may be logically consistent with present quantum mechanics, . . . materialism is not.[4]

The conclusions expressed by Wigner and Peierls in these statements are based on some arguments first made by the great mathematician and physicist John von Neumann[5] and later elaborated upon by the physicists Fritz London and

Edmond Bauer.[6] I will call the claim that quantum theory is incompatible with materialism the London-Bauer thesis.

This thesis, while it has a distinguished pedigree, and has been held and defended by such greatly respected physicists as Wigner and Peierls, is, as might be expected, highly controversial. It is the view which makes the most sense to me, and for which I will present some well-known arguments. At the same time I must caution that it is impossible at present to base any absolutely firm philosophical conclusions upon the structure of quantum theory. There are two reasons for this. First, it is not clear that any of the "interpretations" of quantum theory that have yet been proposed is entirely free of philosophical difficulties. And, second, there is the possibility that quantum theory as we now know it will eventually be superseded. This has long been the hope of many physicists, including Einstein. On the other hand, quantum theory has been around for over seventy-five years now, and there is no evidence that its long sway is coming to an end. Indeed, experts say that superstring theory—which many expect to be the ultimate theory of matter—seems to require that the basic principles of quantum theory *not* be modified. In a recent talk, a noted theorist, after predicting that superstring theory will require profound changes in our notions of space and time, and indeed in almost all our other basic notions about the physical world, said, "At the moment it appears that the only things which may remain unscathed are the fundamental principles of quantum mechanics."[7]

The London-Bauer Argument in Brief

Quantum theory was originally developed to solve certain problems in atomic physics and in the theory of electromagnetic radiation. It dealt originally with atoms and particles of light. It seems strange that it would have anything to say about the mind. But as we have seen, some great physicists, including Wigner and Peierls, have said that it does. In this section I will give a brief outline of their reasons for saying so. In the sections that follow I will flesh out these arguments, and go into greater detail.

The starting point of the London-Bauer argument is the fact that quantum theory is probabilistic. Indeed, all the famous paradoxes of quantum physics go back to this basic fact. In quantum theory, generally speaking, one does not calculate what is going to happen, but rather what might happen and the probabilities of it doing so. (Occasionally, one finds that the probability of something happening is 100 percent. But those are special situations.) This immediately raises a problem: how does a theory that deals only in probabilities get related to a real world in which things definitely happen or do not happen? In calculating probabilities, quantum theory treats events as *hypothetical*. But the real world is made up of *actual* events. What is the link between the two?

Consider an analogy. Jane takes a French exam. A week before the exam, one might say that Jane has a 50 percent chance of getting an A+. After studying a few days, her chances may go up to 80 percent. Maybe after becoming ill her chances decline to 60 percent. But at some point she takes the test and her paper gets graded. Only at that point does her grade cease to be a matter of fluctuating probabilities and become a matter of hard fact. Between the probabilities and the fact, something decisive occurs: the exam and the act of grading. If the grading were never to take place, then the grade would never become a matter of fact, and all the talk of probabilities would become quite meaningless. A "probability" only has meaning as the probability of an outcome; and so there has to be, at some point, a definite outcome in order to talk about probabilities at all. This must also be true of the probabilities calculated in quantum theory.

We are led, then, to ask: When do outcomes happen in quantum physics? What is the analog of "grading" that takes place? When does the hypothetical possibility become recognized as being a fact or not being a fact of the real world?

The answer to these questions that is generally agreed upon is that in quantum theory one goes from probabilities to facts when a "measurement" or "observation" occurs. *The probabilities that are computed in quantum theory are the probabilities of outcomes of measurements.* And this is where the "observer" comes in. The observer is the one who does the measurement and gets a definite result. In classical physics one talked only about physical systems. In quantum physics, however, one talks about *physical systems* on the one hand and *observers* who make measurements or observations of them on the other. This is why quantum physics forces us to confront the question of what this "observer" is, and how mind and matter are related.

What the mathematics of quantum theory describes are physical systems, and these descriptions are always inherently hypothetical or conditional in character. For a particular system, the mathematics describes what might be found by an observer measuring or observing that system, and the relative probabilities of his finding different things. The observer is not part of the system being described. He is effectively an *outside* observer. He *intervenes* in the system by making a measurement or observation of it. Before the observer intervenes, the system is described by the mathematics in terms of multiple, hypothetical possibilities. It is only the observer's intervention that takes one out of the realm of the hypothetical and into the realm of the actual.

So far there is nothing puzzling or paradoxical or problematic in all of this. The problems begin when we start looking at the process of measurement or observation itself. We would like to understand the act of observation scientifically. After all, it involves physical acts performed by a physical being using physical sense organs and, on occasion, physical devices. All of this should be understandable to physicists. Presumably, all of this should be describable using the mathematics of quantum theory. We should be able, one would think, to give a complete mathematical description not only of the experimental devices of the observer but of *the observer himself*—not in practice, perhaps, but at least in principle.

The surprising claim, however, is that this cannot be done. The observer is not totally describable by physics. The reason is the following. If we could describe by the mathematics of quantum theory everything that happened in a measurement from beginning to end—that is, even up to the point where a definite outcome was obtained by the observer—then the mathematics would have to tell us what that definite outcome was. But this cannot be, for the mathematics of quantum theory will generally yield only probabilities. The actual definite result of the observation cannot emerge from the quantum calculation. And that says that something about the process of observation—and something about the observer—eludes the physical description.

We can turn this around and say it the other way. If one attempted to describe the observer in his totality by the mathematics of quantum theory, then he too would become a part of the world of the merely probable. He would step out of the world of facts, so to speak, and into the realm of the hypothetical and conditional. However, the role that the observer plays in quantum physics is precisely to allow an escape from that realm, and consequently if he were to become part of it himself he would not be able to play his assigned role. The theory would only allow one to make statements about what the observer and his equipment *might* find and the probabilities of their finding it, but not what they *actually* find. To find out what they actually find, one would have to have *another* observer come along and observe *them*. In other words, the original observer would have abdicated his role of observer by becoming absorbed into the system being observed. The role of observer would have to be taken up by someone else—someone *outside* of the mathematical description. In the words of Peierls,

> [The transition from hypothetical possibilities to definite outcomes] takes place only outside the "system," which we describe in detail, and belongs to the observation of the "observer," who lies outside of our description.[8]

The upshot is that the observer as such is not describable by the mathematics of quantum theory. Here we begin to see why there might be something about the observer that is non-physical. But what might that something be? Can we isolate it? Whatever it is must be something that makes an observer an observer in the first place. It must be something that allows him to play the role of "finder of fact," so to speak, linking the hypothetical to the factual. And what can that something be but the observer's mind, or at least that part of his mind which judges truth and falsehood?

The point at which the transition is made between the hypothetical and the factual is the point at which the observer judges that one hypothetical possibility did actually occur, and the others did not, that the measurement gave one result and not the others. That is, it is the making of such a judgment of fact that completes the observation or measurement. In traditional terms, the human faculty that makes such judgments is the intellect.

This, in outline, is the reason why, according to some, the logical structure of quantum theory requires that the intellect of the observer lies to some extent beyond the possibility of physical/mathematical description. To summarize the argument briefly, it has these five steps: (1) In quantum theory the fundamental theoretical description is in terms of multiple hypothetical possibilities and their relative probabilities. (2) To relate these hypothetical possibilities to the world of facts, measurements must be made by an "observer." (3) A measurement is complete at the point when the observer has made a judgment of fact, that is, when he comes to *know* that one of the hypothetical possibilities is a fact and the others are not. (4) What makes the observer an observer, therefore, is an act of the intellect. (5) This act cannot be completely described by the physical theory without inconsistency, for the mathematics of the theory gives us only probabilities while the observations yield definite outcomes.

GOING INTO MORE DETAIL

Probabilities

It might seem that some kind of sleight of hand has been done. Somehow we have gotten from the seemingly small fact that quantum theory involves statements about probabilities to a profound conclusion about the human mind. The basic reason we can do this is that the very notion of probability itself is deeply connected with the idea of knowledge. Statements of probability have to do ultimately with someone's level of knowledge or ignorance. So, once probability becomes a fundamental concept in our picture of the world, then *minds that can know* become fundamental too. As Peierls expressed it,

> [t]he quantum mechanical description is in terms of knowledge and knowledge requires *somebody* who knows.[9]

What do I mean by saying that probability has to do with knowledge? The answer, simply, is that if a person knew everything, then for him nothing would be a matter of probabilities, everything would be certain. Normally, we only talk about probabilities when we don't know everything. For example, for those who know the results it is not a matter of probabilities that Lincoln won the presidency of the United States in 1860 (except in the rather trivial sense that the probability is 100 percent). But for a person living in 1859, it did make sense to discuss the election of 1860 in probabilistic terms.

The point can be made even clearer, perhaps, by the following well-known paradox: Three prisoners (call them A, B, and C) are sitting in Fidel Castro's jails under sentence of death. One day, Castro announces that he intends to execute only two of them and set the other free. However, he does not publicly disclose which prisoner is to be spared. Naturally, each prisoner reckons that he has a 1-in-3 chance of surviving. That evening, prisoner A asks the guard the identity of the prisoner who is to be set free. The guard replies, "I am forbidden to tell you; but I will tell you that it is not prisoner C." Prisoner A comforts himself with the thought that his chances of survival have just improved from 1 in 3 to 1 in 2.

The paradox is this: on the one hand it seems quite reasonable for prisoner A to be cheered by the information the guard has given him. It seems that his chances really have improved. And yet, objectively speaking, nothing in the situation has changed. Castro's decision has already been made. Have the probabilities really changed? The answer is that from the point of view of prisoner A they have, but from the point of view of Castro and the guard they have not. Probabilities are a measure of ignorance. On the question of who will be executed, Castro is "in the know," and so for him it is a matter of certainties. But prisoner A is not "in the know," and so he must assign probabilities. If his state of knowledge changes, the probabilities he assigns will naturally change accordingly.

But if it is true that speaking about probabilities involves speaking about someone's level of knowledge or ignorance, and if that means that quantum theory is based on the existence of "knowers," then why isn't the same true of classical physics, i.e., the physics that preceded quantum theory? In classical physics one also uses probabilities extensively. For instance, the whole field of classical statistical mechanics is based on probabilities.

The answer to this is that in classical physics probability does not play the fundamental role that it does in quantum physics. In classical physics one could in principle, given sufficient information, dispense altogether with probabilities. For example, if we lived in a classical, rather than quantum universe, we could still say that the probability of flipping a coin and getting "heads" was 50 percent. However, if one knew exactly how the coin was tossed—its precise position, orientation in space, velocity, and rotational motion at some instant of time, as well as the motion of the air, the elastic properties of the coin, the shape of the surface of the table, and all other relevant variables—one could infallibly predict how the coin toss would come out. And in a classical universe, it would (in principle anyway) be possible to know all of these things with as much exactness as required. In classical physics, then, the use of probabilities is merely a matter of convenience or an accommodation to practical limitations.

That is not so in quantum theory. There one cannot dispense with probabilistic concepts; they are woven into the very fabric of the theory. Even if one had all the information about a physical system which it were possible to have at one time (in the physics jargon, if one knew exactly the values of a "complete set of variables"), one would still not be able to predict everything about the future behavior of that system with certainty.

This deep difference between classical and quantum physics is reflected in the basic quantities that are calculated in the two frameworks. Consider, for example, a simple physical system consisting of a particle of matter moving through space. In the classical treatment, the basic quantities that describe that system are the "coordinates" of the particle. The coordinates are just a set of numbers that tell where the particle is in space at any particular time. As the particle moves, these coordinates change their numerical values. The precise way that they change is governed by a set of equations called the "equations of motion."

In quantum theory, by contrast,[10] the basic quantities that appear in the equations are the so-called "probability amplitudes." These probability amplitudes do not tell where the particle is; rather, they allow one to calculate the relative probabilities that the particle will be found in various places. These probability amplitudes make up something called the "wavefunction" of the particle. This wavefunction evolves in time by a fundamental equation called the Schrödinger equation.

Thus, in classical physics, one computes where the particle actually is from moment to moment, while in quantum physics one computes probability amplitudes for where the particle is. I have used the example of a particle moving through space. But everything I have said applies to physical systems of all kinds. In classical physics, any physical system, no matter how complicated, can be completely specified by a set of coordinates. The coordinates need not be the positions of particles. They could be the strengths of electromagnetic fields, say, or the curvature of space-time. In quantum theory one would instead compute the probability amplitudes that electromagnetic fields have a certain strength or that space-time is curved in a certain way.

There is, then, a profound difference in the way probability enters in the two frameworks. In the classical framework, the use of probability is not in principle necessary, whereas in the quantum framework it is. In quantum theory the probability amplitudes are at the very heart of the mathematical description of physical reality.

Outcomes

The next step in the chain of argument is to ask what the probabilities in quantum theory are probabilities *of*. And, of course, the answer must be that they are probabilities of *outcomes* of some kind. However, that raises the question of how and when the outcomes happen.

This is a somewhat embarrassing question. The problem is that the equations of quantum theory do not, in general, yield definite outcomes, they yield only probabilities. Let us consider a simple example: a radioactive nucleus that has a half-life of one hour. The Schrödinger equation for that nucleus allows one to calculate the probability that it has "decayed" (that is, undergone radioactive disintegration) by any given time. Suppose that to start with one determines that the nucleus is still intact at 12:00 noon. That obviously means that at 12:00 noon there is a 0 percent probability that the nucleus has decayed. But by 1:00 P.M. there is a 50 percent probability that it has decayed (because the nucleus has a half-life of one hour). In other words, the "probability amplitudes" have evolved over the course of that hour. They have evolved in a precise way governed by the Schrödinger equation. By 2:00 P.M., the probability that the nucleus has decayed is 75 percent. The Schrödinger equation allows one to compute the probability at any time, of course, not just every hour on the hour. For instance, at 3:45 P.M. there will be a 92.568 percent chance that the nucleus has decayed. However, at no time after 12:00 noon will the wavefunction evolving by the Schrödinger equation describe a situation in which the nucleus has definitely decayed, or a situation in which it has definitely not decayed.[11]

When, then, do these probabilities get replaced by definite outcomes? The traditional answer is that this happens when a "measurement" or "observation" is made. As we have said, the "outcome" is always the outcome of a measurement.

In our example, the mathematics says that at 3:45 P.M. there is a 92.568 percent chance of the nucleus having decayed and a 7.432 percent chance of it still being there. However, if one were actually to take a look at 3:45 P.M., one result or the other would be found. One is not going to find of this particular nucleus that it is 92.568 percent decayed. It has either disintegrated or it hasn't. So how does one test the prediction of the theory that gave the probability as 92.568 percent? By looking at a very large statistical sample of nuclei of the same type: after three hours and forty-five minutes it should be found that 92.568 percent of them have disintegrated. But an observation of one nucleus designed to see whether it has disintegrated must yield one outcome or the other.

The crucial point is that only by talking about measurements made on systems, and the outcomes of those measurements, does it seem to be possible to make sense of the mathematical formalism of quantum theory. However, this leads immediately to the most critical issue of all. What are these "measurements," and who is it that performs them? A consideration of these questions will force us to deal with the observer. It is here that mind will come into the discussion.

Measurements

In quantum theory, as traditionally formulated, there are "systems" and "observers." Or rather, in any particular case, there is *the* system and *the* observer. The observer makes measurements of the system. As long as the system is undisturbed by external influences (that is, as long as it is "isolated") its wavefunction—which is to say, its probability amplitudes—will evolve in time by the Schrödinger equation. We saw this in the example of the radioactive nucleus, where the decay probability at 1:00 P.M. was 50 percent, at 2:00 P.M. was 75 percent, and so on. However, when a measurement is made of the system the observer must obtain a definite outcome. Suddenly, the probability for the outcome that is actually obtained is no longer what the mathematics said it was just before the measurement, but jumps to 100 percent. And the probabilities for all the alternative outcomes, the ones that did not occur, fall to 0 percent.

There is nothing paradoxical about this. It is common sense. To use an earlier example, before the election of 1860, there may have been (according to someone's calculation) a 60 percent chance of Lincoln winning, but after the votes were tallied, the probability jumped to 100 percent.

What this means in the case of quantum theory is that after a measurement is completed, i.e., after the observer has learned the outcome, the probabilities as far as the observer is concerned are no longer what they were just before he learned the outcome. Consequently, after a measurement the old wavefunction must be replaced by a new wavefunction that reflects the observer's newly acquired knowledge. This change or replacement of the wavefunction is traditionally called "the collapse of the wavefunction." It is essentially the point at which the probabilities get turned into certainties.

According to the traditional analysis, which is due to von Neumann, the wavefunction and the probability amplitudes it contains change in two radically different ways: (1) the Schrödinger evolution of the wavefunction, and (2) the collapse of the wavefunction.

First, the Schrödinger evolution of the wavefunction:

> In between observations, if a system is undisturbed, its wavefunction evolves in a continuous way which is governed by the Schrödinger equation. This evolution is smooth and predictable, in the sense that if the probability amplitudes are specified at one time, their values are computable, in principle, at later times.

Second, the collapse of the wavefunction:

> When an observer who is outside the system performs some measurement or observation of it to determine one of its properties, then some probability amplitude (that for the actual outcome) jumps to 100 percent, and the others

(for the outcomes that were not realized) jump to 0 percent. This collapse is sudden, but more importantly it is—unlike the Schrödinger evolution—*un*predictable. The collapse is unpredictable since which probability jumps to 100 percent and which jump to 0 percent depends on the actual outcome of the measurement or observation, and that is not predictable in advance.

To repeat once more the heart of the argument: If the "collapse" of the wave-function to a definite result were computable by the Schrödinger equation, then that definite result would be computable in advance. But that is not possible, since the Schrödinger equation only gives probabilities.

(One thing that should be noted is that while a measurement always makes some quantities or properties of the system more certain—namely those that are being measured—it always makes others more uncertain. This is what Heisenberg's Uncertainty Principle implies. There are properties of every physical system that are said to be "conjugate" to each other: the more certainly one is known, the less certainly the other is known. For example, if one measures the position of a particle in space, its position becomes better known, but its momentum becomes less well known. But could one just keep making more and more measurements of a system until everything about it is known with certainty, and then perhaps dispense with all these "probability amplitudes"? The answer is no. Any measurement will produce greater certainty about some aspects of the system, but other aspects will become more uncertain. Therefore, probability is an ineradicable feature of quantum theory.)

The Observer

Having this framework in mind, we are now in a position to say something about what goes on during a measurement, and, in particular, something about the observer.

Who or what is this observer? Can the observer be something physical, like the system that he, she, or it is observing or making measurements on? For example, if the system is a radioactive nucleus, can the observer be a Geiger counter? Or if the system is a particle moving in space, can the observer be a camera that takes a picture of where that particle is? The surprising answer is no—at least according to von Neumann, London, Bauer, Wigner, Peierls, and those who accept their reasoning.[12] For one runs into logical difficulties if one imagines the observer to be something *entirely* physical, like a camera, or a Geiger counter. To be more precise, one runs into problems if one tries to give a complete mathematical/physical description of the entire process whereby the observer obtains an outcome to a measurement.

It is easiest to see why this is so by considering a simple example. Let us think about a physical system consisting of a particle of matter moving through space.

And let us suppose that a camera takes a snapshot of that particle and records its position on film. The first thing to note is that the camera and the film are themselves just physical systems made up of particles of matter—atoms. It is therefore possible to describe the behavior of the camera and film using the equations of quantum theory, just as one can describe the behavior of the particle that they are photographing. In particular, it is possible, in principle, to write down an enormously complicated wavefunction that describes the whole complex meta-system (as I shall call it) that consists of the particle, the camera, and the film. That wavefunction will evolve in accordance with a Schrödinger equation. And such an evolution is smooth, predictable, and yields only probability amplitudes, as we have said.

Now certainly the camera does "capture on film" the actual motion of the particle. To be more precise, what the particle does and what the camera records on film are *correlated*: if the particle is in one place in space (call it A) when the snapshot is taken, its image will appear at some corresponding place on the film (call it A'); while if the particle is at some different place in space (call it B), its image will appear in some different place on the film (call it B'). *Nevertheless, the camera taking the snapshot does not produce a definite outcome.* It cannot, for the wavefunction that describes what happens to the meta-system of particle, camera, and film only contains probability amplitudes. These amplitudes will only say that there is some probability (say P_A) that the particle is at A and that its photographic image is at A', some other probability (say P_B) that the particle is at B and that its photographic image is at B', and so on. But it won't tell us *which* of those cases is actually realized.

The problem is that by bringing the camera and film inside "the system" (which we have now called the meta-system) they are no longer outside where they can play the role of observer. By expanding the mathematical description to include the camera and film, the camera and film become part of the continuous Schrödinger evolution of a wavefunction. One is then trapped again in the realm of probabilities: there is nothing outside the new, larger wavefunction to collapse it. To quote the words of Peierls again, this time more fully: "[T]he 'collapse' of the wavefunction always takes place only outside the 'system,' which we describe in detail, and belongs to the 'observation' of the observer, which is not part of our description."[13]

In short, the observer cannot be considered part of the system that is being physically described and remain the observer of it. Just as you cannot be in the movie and watch it at the same time, you cannot be entirely part of the system and observe it too. You cannot be described completely by the wavefunction and also collapse it. In traditional quantum theory one is led to the following fundamental conclusion: *The mathematical descriptions of the physical world given to us by quantum theory presuppose the existence of observers who lie outside those mathematical descriptions.*

The Line between System and Observer

We have said that the observer lies outside of the mathematical description given by quantum theory. But does this mean that nothing about the observer can be physically described? That would be absurd. After all, human observers have hands and eyes and other parts with which they make observations of the world around them; and these parts are certainly physical. Hands and eyes are made up of the same stuff as everything else: molecules, atoms, and subatomic particles. So what exactly is our physical description unable to reach?

To answer this, let us consider what goes on during a measurement. In particular, consider what happens when a human being uses certain equipment to observe the motion of a particle in space. One way he may observe the particle is by recording the track it makes in a "bubble chamber" using a camera, and subsequently examining the camera's photographic plate. Normally, in such a situation, one would think of the particle itself as the system being observed, whereas the bubble chamber, camera, and photographic plate would be regarded as extensions of the observer. However, since bubble chambers, cameras, and photographic plates are certainly physical objects, there is no reason in principle why they could not be considered part of the physical system being observed. In that case, the observer would consist of just the human experimenter himself, who does his observation by looking at the photographic plate.

But, then again, the human experimenter makes use of his sensory organs to examine the images on the photographic plate; and these sensory organs are certainly also capable of being understood physically. There is no fundamental difference between the eyeballs of the experimenter and cameras. So there is nothing to prevent one from considering the experimenter's eyeballs as being part of "the system" too, along with the particle, camera, and photographic plate, while the rest of the experimenter is considered "the observer." We see, then, that where we place the boundary line distinguishing system from observer is to a great extent arbitrary.

Suppose we place the boundary line so that not only the bubble chamber, camera, photographic plate, and eyeballs, but also retinas, optic nerves, and the visual centers of the experimenter's brain are considered part of the physical system. In that case one has a comprehensive physical description of everything that happens when the particle being observed moves through space; when it causes a track in the bubble chamber; when light reflected off that track enters the camera and is focused by its lens system onto a photographic plate; when light from the image on that plate enters the experimenter's eyeballs and is focused on his retinas; when the light that impinges on the retinas causes complicated chemical processes in the retinal cells; when those reactions produce electrical impulses that travel down the experimenter's optic nerves, past the "optical chiasma," into the "lateral geniculate bodies," and thence into the left and right visual cortex of the brain. One could certainly imagine such a comprehensive physical description—in principle anyway.

At the other extreme one can place the boundary line so that only the particle moving through space is considered the system and everything else is the observer. However, although the boundary between system and observer can be shifted very far in one direction or the other, it cannot be erased altogether. The human observer cannot be brought over *entirely* onto the "system" side of the line. For if he is, then his own behavior will be described by some wavefunction in terms of multiple, merely hypothetical possibilities. He can no longer then be the observer whose act of measurement or observation reduces the hypothetical to the actual and the multiple possibilities to one measured outcome. In Wigner's words,

> [e]ven though the dividing line between the observer, whose consciousness is being affected, and the observed physical object can be shifted towards the one or the other to a considerable degree, it cannot be eliminated.[14]

The Mind of the Observer

In shifting the boundary between physical system and observer back and forth, something must always remain on the observer's side of the boundary. What is that something? It is hard to avoid the conclusion that it is the consciousness or mind of the observer, or at least some part of his mind. In fact, this is what Wigner is suggesting when he refers to "the observer, whose consciousness is being affected." Why the mind? Because it is the mind that *knows*. It is when the observer knows the outcome of the measurement that he can say that this outcome now has a 100 percent probability (because it is known to have happened) and the other possible outcomes have 0 percent probability (because they are known not to have happened). To quote Peierls again:

> [T]he moment at which you can throw away one possibility and keep only the other is when you finally become *conscious* of the fact that the experiment has given one result. . . . You see, the quantum mechanical description is in terms of knowledge, and knowledge requires *somebody* who knows.[15]

What is suggested by the foregoing analysis (which is essentially that of von Neumann) is that there is something about the mind of the observer that is not entirely reducible to description in terms of wavefunctions, and which, so to speak, consummates the act of observation and causes the wavefunction to collapse. Hermann Weyl, reflecting upon the implications of quantum theory, wrote,

We may say that there exists a world, causally closed and determined by precise laws, but in order that I, the observing person, may come into contact
with its actual existence, it must open itself to me. The connection between
that abstract world beyond and the one I directly perceive is necessarily of a
statistical nature.[16]

The traditional interpretation thus seems logically to lead to the conclusion
that the mind of the "observing person," to use Weyl's phrase, plays a fundamental role in quantum theory. In the passage previously quoted from Wigner's
classic essay, he went on to say,

It may be premature to imagine that the present philosophy of quantum
mechanics will remain a permanent feature of future physical theories; it will
remain remarkable, in whatever way our future concepts may develop, that
the very study of the external world led to the conclusion that the content of
the consciousness is an ultimate reality.[17]

(At this point I should put in a technical remark for experts. The foregoing
analysis seems to lean heavily on the use of wavefunctions. But experts will realize
that at the stage of the measurement process when macroscopic objects—such
as cameras or eyeballs—become involved, what is happening can no longer be
described in practice by a wavefunction. One must really use the density matrix
formalism. Moreover, at the stage when macroscopic objects begin to be affected, the parts of the wavefunction—or density matrix—that represent different
possible outcomes "decohere" from each other. That, however, does not affect
the central point of the foregoing line of argument. It remains the case that the
evolution given by the equations of quantum theory, whether one is speaking of
a wavefunction or of a density matrix, does not tell *which* outcome is actually
going to happen. The actual "collapse" is not merely a matter of decoherence,
it must result in a definite actual outcome, and therefore cannot be given by the
equations of standard quantum theory.)[18]

The idea that the "mind," or "consciousness," or "knowledge" play a fundamental role in quantum theory was and is accepted by a significant number of
physicists, but they do not all have precisely the same ideas about what role the
mind plays and how it plays it. That the mind is a fundamental reality that is not
reducible to physical description was clearly stated by Fritz London and Edmond
Bauer, and later defended by Sir Rudolf Peierls[19] and by Eugene Wigner,[20] as
we have just seen. Henry Stapp[21] of the University of California at Berkeley and
Euan Squires[22] of the University of Durham in England are the physicists who
have recently argued most forcefully for this point of view. Each has proposed
interesting theories about the nature and role of the mind that are based on the
ideas of quantum theory.

There are others whose views on quantum theory are significantly different, but who concede that "consciousness" may play a significant role of some kind. For example, John A. Wheeler, who helped develop the alternative "many-worlds" interpretation of quantum theory, later returned to the traditional interpretation. He said in a recent interview on the subject, "I would not like to put the stress on consciousness even though that is a significant element in this story."[23]

Some physicists express similar ideas by saying that the wavefunction represents the "knowledge of the observer." For example, Heisenberg himself wrote that the mathematics of quantum theory "represents no longer the behavior of elementary particles, but rather our knowledge of this behavior."[24]

A noncommittal attitude was expressed by John Bell, who invented the mathematical relation called Bell's inequality, which plays a major part in modern discussions of quantum theory. Bell said, "I can see the logic of people who say that it [analysis of the process of measurement] goes in the direction of showing that mind is essential. It's a hypothesis we can certainly explore, but I don't know that it's the only one."[25]

Though it seems to follow logically from the traditional interpretation of quantum theory, the idea that the mind plays a role is highly controversial, and is looked at with great suspicion by many, if not most, physicists. As Euan Squires observed, "It is probably fair to say that most members of the physics community would reject [these] ideas. . . . [However], their reasons would be based more on prejudice than on sound argument, and the proportion of those who reject it would be much smaller if we considered only those who had actually thought carefully about the problems of quantum theory."[26]

Is the Traditional Interpretation Absurd?

At first sight it would appear that the traditional interpretation of quantum theory, at least the implications drawn from it by such physicists as London, Bauer, Wigner, and Peierls, gives an absurd picture of the world. The most disturbing feature of this picture is that it seems to say that we can change the external world just by thinking about it. However, this is a false impression, as I will now explain.

It is natural to think of the wavefunction as representing the state of the physical world. If that were so, then the collapse of the wavefunction would represent a sudden change in the external world brought about by the observer's measurement of it. And since that measurement is consummated (according to von Neumann's analysis) by the observer becoming conscious of its outcome, it would seem, indeed, that in quantum theory "thinking makes it so."

The fatal misstep is in thinking of the wavefunction as representing the actual state of the physical world in some unqualified sense. Instead, it is necessary, if one is to avoid obvious absurdities, to understand the wavefunction as representing someone's state of knowledge of the physical world. This is in fact exactly the language Heisenberg used, as we saw in the previous section: he said that the mathematics of quantum theory "represents no longer the behavior of elementary particles, but rather our knowledge of this behavior." And Peierls says, "quantum mechanics is capable only of describing the knowledge of an observer outside the system."[27]

If the wavefunction represents the observer's state of knowledge of the system, then there is nothing objectionable or mysterious in the fact that a change in his knowledge results in a change in the wavefunction. On the contrary, it would be absurd to say otherwise.

This recognition that the wavefunction is someone's state of knowledge lets us out of some of the more blatant paradoxes of quantum theory. One set of paradoxes concerns the question of what happens if there are several observers of the same system. In such a case, which observer collapses the wavefunction? The most famous paradox of this type is called the "Wigner's friend paradox."[28] In this paradox, Wigner asked what would happen if he himself were the observer of some physical system, but made use of a friend as an assistant. Suppose Wigner left the laboratory for some period of time and had his friend keep watch over a piece of apparatus set up to detect (say) the radioactive decay of a nucleus. And suppose that the friend sees that the nucleus decays sometime during Wigner's absence. Wigner could choose to regard his friend as being in effect merely part of the experimental apparatus. The actual measurement, in this case, would not be complete, and the wavefunction would not collapse, until the knowledge of the outcome entered Wigner's consciousness (since he is the "observer"), and this does not happen until Wigner comes back to the laboratory and asks his friend what happened. On the other hand, Wigner's friend could just as well regard himself as being the observer, in which case the measurement is complete and the wavefunction collapses when the result enters his own (the friend's) consciousness, hours before Wigner learns of it. It would seem that Wigner and his friend have an equal claim to be the "observer," so when does the wavefunction really collapse?

The resolution of this paradox is that, since there are two states of knowledge involved that do not coincide, namely Wigner's and his friend's, there are really two distinct wavefunctions that one is talking about. There is the wavefunction that encapsulates the state of Wigner's knowledge of the system, and a wavefunction that encapsulates the state of his friend's knowledge of the same system. (One need not think of wavefunctions as being somehow personal possessions of different people. It is better to think of a wavefunction as containing all that it is possible to assert about a system given a certain set of prior observations. Then if two people have the same information about a system they would employ the same wavefunction to describe it.)

Distinguishing between different observers' wavefunctions for the same system is a necessary consequence of regarding wavefunctions as representing states of knowledge, since different observers can have different states of knowledge. The danger that many see in this conception of the wavefunction is that it might lead to a subjectivism in which it is no longer possible to speak of a unique objective reality existing independently of human perceptions of it. But it is not clear that one has to fall into that trap; at least, the more blatant kinds of subjectivism are certainly avoided. In particular, it can be shown that if several observers compare notes about the same set of physical facts they will have to be in agreement with each other—unless, of course, one or more of them makes an error of some sort. If Wigner asks his friend what happened, and then sees for himself, he will find the same thing (unless his friend made a mistake or lied). So certainly in quantum theory different observers can study the same reality, and if they do so properly they will reach consistent conclusions about it.

Nevertheless, it is true that one has gone from a straightforward account (as in classical physics) of "what is happening" to a patchwork of what different observers are in a position to know about what is happening. This raises the interesting question of whether we can talk about what happens if no one ever observes it. What if the universe had never given rise (and never were to give rise) to sentient beings such as ourselves? What then would the wavefunctions refer to? What would the laws of physics mean without observers?

From a narrowly scientific point of view, such questions can be safely ignored, in fact must be. What science tries to do is explain the results of observations of the physical world. It lies outside empirical science to ask what would happen in the absence of any observations. And certainly, whatever conclusions we reached on that subject could never be tested. Nevertheless, these are important questions for philosophy; and I share the view of many writers that these questions constitute the real challenge for the traditional interpretation of quantum theory.

25 Alternatives to Traditional Quantum Theory

Everything that I have said up to this point about the implications of quantum theory is based on the fundamental idea of the "collapse of the wavefunction" when observations are made. This idea is not the suggestion of a radical fringe of physicists. It is the traditional way of looking at quantum theory, and is so mainstream that it is, as I have said, sometimes called the "orthodox" view. However, if one begins to take this view seriously, and ask where it leads, one comes to rather startling, and to some people shocking, conclusions.

There are two ways that physicists who dislike the traditional interpretation of quantum theory deal with the situation. Most of them respond in the way suggested by the philosopher Hume in another context: with "carelessness and inattention."[1] In other words, they just ignore the issues. This is perfectly reasonable. For the practical business of doing physics—of actually calculating what the results of experiments will be—it is not necessary to worry about the philosophical implications of the theory. However, it is not only practical absorption in their business that leads physicists to adopt this attitude. Many physicists are made uneasy and impatient by discussions of the philosophical implications of quantum theory. Partly this is because they know that these implications are strange, and they would rather not be forced to think about them or discuss them, like the doings of some embarrassing relative; and partly it is because physicists look askance at anything that smacks of metaphysics. The present-day scientific community is heavily imbued with positivist and skeptical attitudes, which lead many to dismiss all philosophical inquiry as useless playing with words.

The other reaction has been to seek a way of modifying quantum theory or of reinterpreting it so that the "observer" does not play a fundamental role. None of these reinterpretations at present commands the allegiance of more than a

245

minority of physicists.[2] I will discuss them briefly in this chapter. In my view, the alternatives to the traditional interpretation of quantum theory all either fail to come to grips with the philosophical issues, or are unsatisfactory for some other reason.

MODIFYING QUANTUM THEORY

In the history of science, when a theory gives rise to seemingly insuperable problems it is often a sign that a "paradigm shift" is required, to use the term made fashionable by Thomas Kuhn. Indeed, quantum theory itself was developed partly in response to inconsistencies and paradoxes that emerged within the framework of classical physics. It is quite natural, therefore, that many physicists have regarded the paradoxical features of quantum theory as an indication that the theory is incomplete or in need of modification. This was the view of Einstein.

Another reason that many theorists have anticipated the need to modify quantum theory is that Einstein's theory of gravity (i.e., general relativity) seemed hard to reconcile with quantum principles. Attempts to treat the force of gravity in accordance with the basic postulates of quantum theory led to mathematical absurdities. For example, when one calculates the "quantum corrections" to the predictions of Einsteinian gravity, they generally come out to be infinite. On account of these difficulties, it was often said that a synthesis of general relativity and quantum theory would only be possible in a framework in which both theories would be changed in some profound way. Recent developments have somewhat undercut this belief, however. Superstring theory appears to be a perfectly consistent quantum theory that incorporates Einsteinian gravity; and in superstring theory the basic principles of quantum theory seem to be left untouched.[3] Even if superstring theory turns out to be a blind alley, it will at least have shown that Einsteinian gravity and standard quantum theory can be reconciled with each other.

Be that as it may, the search for ways to change quantum theory so as to rid it of its paradoxical features has gone on for many decades. One of the oldest ideas goes by the name "hidden variables theory." The hidden variables idea is essentially an attempt to go back to classical concepts. We have seen that the root of the paradoxes and puzzles of quantum theory is the fact that it is inherently probabilistic in nature. Classical physics by contrast is not inherently probabilistic. In classical physics it is only because one often in practice does not have access to all the relevant information for predicting a system's future behavior that one is forced to use probabilities. However, that is a mere practical necessity, not an absolute one. In classical physics there is no theoretical bar to acquiring all the information needed in order to dispense with probabilities altogether. The idea of hidden variables theory is that the need for probabilities in quantum theory is really ultimately of the same kind. That is, it is suggested that it only arises from the fact that we do not have practical access to all the information about the systems we study. The information we lack is contained in the "hidden variables."

Einstein took the view that the probabilities in quantum theory merely reflect the existence of hidden variables. However, this idea has now been largely discredited. In 1965, the physicist John Bell showed that in certain situations one could distinguish the probabilities that are predicted by quantum theory from those that could arise as a result of hidden variables. The latter would have to satisfy a certain mathematical relationship, which is now called "Bell's inequality," whereas the former would not have to. In 1982, experiments performed by Alain Aspect and his collaborators found that in certain physical systems Bell's inequality was definitely violated. This is regarded as decisive evidence against the hidden variables alternative to quantum theory.

Another and more subtle attempt to return to more classical concepts goes by the name of "pilot wave theory." The basic idea here was proposed by Louis de Broglie, one of the founders of quantum theory, and has since been developed by David Bohm and many other researchers.[4] In pilot wave theory there is still a wavefunction, and it still satisfies the same Schrödinger equation as in standard quantum theory. However, the wavefunction plays a fundamentally different role. It is no longer thought of as being made up of "probability amplitudes." Rather, it is simply a conventional force field (somewhat like an electric or magnetic field, say) that influences the behavior of the physical system. The physical system itself is described by a set of coordinates, much as in classical physics. These coordinates have their own "equations of motion," again much as in classical physics. The wavefunction field appears in these equations of motion and acts to "pilot" the motion of the system, as it were.

Pilot wave theory is thus a kind of hybrid of classical and quantum ideas. Its basic structure is classical, but some of its equations are the same as the equations of standard quantum theory. Because it is basically a classical theory, many of the paradoxical features of standard quantum theory simply do not arise.

It has been shown that pilot wave theory can reproduce the results of standard quantum theory—that is, give the same predictions for the results of experiments—when applied to a wide range of physical systems. However, most physicists do not find pilot wave theory very plausible or appealing. There are several reasons for this. One reason is technical, and may be temporary. As currently formulated, pilot wave theory treats time and space very differently from each other. It thus seems more compatible with a Newtonian conception of time than with the theory of relativity. In fact, it has not yet been shown that pilot wave theory can be applied to "relativistic" systems (i.e., systems where velocities near the speed of light are important).

A more basic objection is that pilot wave theory seems quite artificial and complicated compared to standard quantum theory. Pilot wave theory has a two-tiered structure: on top of the structure of a wavefunction evolving in accordance with a Schrödinger equation, there is added an elaborate superstructure involving classical-type coordinates evolving in accordance with a new set of equations that do not appear in standard quantum theory. This greater complexity might be worth it if pilot wave theory were more powerful or successful as a physical theory, but it is not. In fact, it seems less powerful, since it has yet to be shown that it can reproduce all of the successes of standard quantum theory.

Another aspect of pilot wave theory that makes it appear quite suspect to many physicists is the way that the two tiers of the theory are related to each other. The wavefunction or pilot wave acts upon and influences the coordinates that describe the behavior of the system, but those coordinates do not act back upon or influence the pilot wave. This is not what one is used to in physics. Usually if A influences B, then B also influences A. For example, electromagnetic fields exert forces on electric charges, and electric charges act as sources of electromagnetic fields. In fact, so pervasive is this reciprocity in physics that Einstein wrote, "[It] is contrary to the mode of thinking in science to conceive of a thing . . . which acts itself, but which cannot be acted upon."[5] For all these reasons, pilot wave theory is regarded by most physicists as a very interesting but probably misguided idea.

Hidden variables and pilot wave theory do not exhaust the ways that physicists have tried to modify quantum theory in order to sidestep its philosophical dilemmas. Another approach that has been tried is to change the mathematical form of the Schrödinger equation in such a way as to make it possible for it to describe the mysterious-seeming "collapse of the wavefunction." A basic challenge for this approach is how to build the unpredictability of the collapse into the new equation.

REINTERPRETING QUANTUM THEORY: THE "MANY-WORLDS" IDEA

The foregoing ideas assume that the basic mathematical structure of quantum theory needs to be changed. If, however, that structure is here to stay—as all present indications suggest—then the only hope for a change is through a change in the way that the mathematics is interpreted.

In 1957, a new interpretation of quantum theory was proposed by Hugh Everett[6] that seems to some physicists to be much simpler and more satisfying than the traditional or "orthodox" view. As we have seen, the source of much of the dissatisfaction with the traditional view is the strange phenomenon called the collapse of the wavefunction. Everett had the brilliantly simple idea that all the knotty questions raised by the collapse of the wavefunction could be avoided simply by saying that the collapse never happens. Let us see where this leads.

As explained in the previous chapter, the collapse of the wavefunction is the point at which the probabilities of quantum theory get converted into definite outcomes. It was argued that when a measurement is made on a physical system, that measurement must have a definite and unique result. If, for example, one looks to see whether a radioactive nucleus has decayed or not, the result must be either "yes" or "no." If it is yes, then at that point the probability that it has decayed is 100 percent and the probability that it has not is 0 percent, whatever those probabilities may have been just before the measurement. Thus, a measurement involves a sudden change of the "probability amplitudes" in the wavefunction. This sudden change is the notorious "collapse."

If we say, as Everett did, that the wavefunction of a system never collapses, then when do definite and unique outcomes happen? The answer is that they don't. Let us consider again the example of the radioactive nucleus. Just before the observer measures to see whether the nucleus has decayed, the wavefunction may say that the probability amplitude for it to have done so has the value A, and the probability amplitude for it not to have done so has the value B. (Assume that neither A nor B is zero.) According to the Everett interpretation, just after the observer makes his measurement the probability amplitudes are still A and B (or very close to those values)—they do not jump. Even after the measurement, neither outcome has a probability of 100 percent. How is that to be understood?

In the Everett interpretation, just after the measurement there is a probability amplitude A for the nucleus to have decayed *and for the observer to see that the nucleus has decayed*, and a probability amplitude B for the nucleus still to be there *and for the observer to see that the nucleus is still there*. Both outcomes happen. Both are equally "real." Reality has two "branches," as it were. In one branch the observer sees that the nucleus has decayed, while in the other branch the observer sees that it has not. The probability amplitudes tell one the relative "thickness" of those branches, so to speak.

This is why the Everett interpretation of quantum theory is usually called the "many-worlds interpretation." All the states of affairs that have non-zero probability amplitudes in the wavefunction are regarded as co-existing and as being equally real, even after a measurement. To say I performed a measurement and "know" the result of it is, strictly speaking, wrong in the Everett interpretation. What is supposed to happen, rather, is that as a result of making a measurement my consciousness splits up: one version of "me" experiences one outcome while the other versions of "me" experience the other outcomes. Each version of me— being unaware of the other versions—naturally thinks of the outcome he experiences as the one and only "real" outcome.

In the many-worlds interpretation, it is an inescapable fact that reality is infinitely subdivided, and that each human being exists in not one, or even a few, but in an infinite number of copies, with infinitely various life experiences. In some branches of reality you are reading this page, in other branches you may be lying on a beach somewhere, or sleeping in your bed, or dead.

This may seem crazy; and, in fact, it seems crazy to physicists too, even to the advocates of the many-worlds viewpoint. These advocates point out, however—and quite rightly—that this kind of co-existence of apparently contradictory possibilities is also a feature of the traditional interpretation of quantum theory, though in a less extreme form. This is illustrated by the well-known example of the "double-slit experiment." This experiment shows that an electron, say, really can be, in a certain sense, in two places at the same time. It is one thing, however, to say that an electron, or even a macroscopic inanimate object, can have such a divided existence. But to say that a rational being can is a more radical proposal. Most physicists are just as reluctant to accept the many-worlds interpretation as they are to fully accept the implications of the traditional interpretation of quantum theory.

One argument that is sometimes made in favor of the many-worlds interpretation is that it is forced upon one by any attempt to describe the entire universe by the laws of physics. If the "system" one is studying is the whole universe, then there cannot be any observer to collapse the wavefunction, since by definition the observer has to stand outside the system and make measurements of it. But no observer can stand outside the entire universe and make measurements of it. So the wavefunction that describes the entire universe can never collapse. Consequently, the wavefunction that describes the entire universe must be interpreted in a many-worlds way.

I believe that this line of argument is based on a verbal confusion, and in particular on an equivocal use of the word *outside*. All that is required in the traditional interpretation of quantum theory is that the observer of a system lie (at least in part) outside the system in an informational sense, not in a geometrical sense. That is, what is necessary is that a complete specification of the values of all the variables (or "coordinates") of the system does not give a complete description of the observer. If a human being is observing the physical universe, he indeed lies geometrically inside the universe that he is studying, in the sense that his body is contained within it. But if his intellect is not something purely physical, then even a complete specification of the values of all the variables of the physical universe would not completely describe his mind. To the extent that his mind is not entirely physical, it would indeed lie "outside" the physical universe.

The great advantage of the many-worlds idea, as seen by its advocates, is its simplicity. Everything, including everything that goes on during measurements, can be described by a wavefunction that evolves at all times simply in accordance with the Schrödinger equation. Gone is all that bizarre business of wavefunction collapse. Gone with it is the fundamental importance of the observer. The many-worlds interpretation seems to slice through all the paradoxes of quantum theory like Alexander's sword through the Gordian knot.

While the many-worlds interpretation is simple, however, it may be too simple. With the bathwater of the collapse of the wavefunction, it may have thrown out the baby of probability. The problem has to do with how probabilities are calculated in quantum theory. I have said that the wavefunction consists of a set of numbers that are called the probability amplitudes. What I have neglected to mention until now, however, is the crucial point that these probability amplitudes are not themselves the actual probabilities of outcomes of measurements. Rather, they are related to those probabilities by a precise mathematical rule, called the "probability rule" or "measurement principle," which is as follows. *The actual probability for a particular outcome of a measurement is given by the "absolute square" of the probability amplitude for that outcome.* (If the probability amplitudes were ordinary numbers, one would just take the ordinary square—the square of a number is the number multiplied by itself. But the probability amplitudes are actually what are called "complex numbers," and the "absolute square" is a way of multiplying a complex number by itself.) For example, in standard quantum theory the probability amplitude that a nucleus has decayed might at a certain time have the value 0.60. That means that if a measurement is performed at that time its chance of showing that the nucleus has decayed is $0.60 \times 0.60 = 0.36$ or 36 percent.

The important point is that in standard quantum theory, just as there are two kinds of processes (the evolution of isolated physical systems, and measurements on those systems), and two kinds of change in the wavefunction (the Schrödinger evolution and the collapse), so there are two mathematical rules: the Schrödinger equation and the probability rule.

The problem with the many-worlds interpretation is that it gets rid altogether of measurement as a distinct kind of event. There is therefore no point at which the probability rule can come into play as a distinct principle. In the many-worlds interpretation one is left with only the Schrödinger evolution of wavefunctions. The trouble with that is that it is questionable whether, with only the Schrödinger equation at one's disposal, there is any way to rigorously deduce the connection between the probability amplitudes in the wavefunction and actual real-life probabilities. Various people have claimed to have done this, but these claims are disputed.[7] It has been argued by some authors that it is in fact impossible to rigorously prove the probability rule in the many-worlds interpretation of quantum theory.[8] If that is indeed the case, it would be fatal to the many-worlds idea.

One point that should be emphasized is that it is in practice impossible on the basis of any experiment to decide between the many-worlds interpretation of quantum theory and the traditional interpretation (supposing, of course, that the potentially fatal problem just discussed can be resolved, and that the many-worlds interpretation makes sense at all). The other branches of reality that are supposed to exist in the many-worlds idea are unobservable. It can be shown that they "decohere" from us.

The many-worlds idea could conceivably turn out to be a viable interpretation of quantum theory. If so, it allows the materialist to escape from the apparent implications of quantum theory that I discussed in chapter 24. The statement of Wigner that materialism is not consistent with quantum theory does not apply to the many-worlds interpretation. The price to be paid for eliminating the observer and the observer's mind from the picture in this way, however, is the postulating of an infinite number of branches of reality, with an infinite number of versions of every person, that are completely unobservable to us.

(I should make a technical point. A great deal of work has been done in recent years on the idea of "decoherence." It has been claimed that decoherence is in itself a way of interpreting quantum theory that resolves all the philosophical issues raised by it. This is not the case, however. Decoherence is not a new interpretation of quantum theory; it is a phenomenon that happens within quantum theory however it is interpreted. While the fact of decoherence is important for making sense of quantum theory by showing that it does not contradict our everyday experience, it leaves one with the same set of alternative interpretations: either the traditional collapse of the wavefunction or the many-worlds picture.)

26 Is a Pattern Emerging?

Religion is sometimes attacked by materialists as a realm of make-believe and speculation in which untestable assertions are made about things that cannot be observed. It is true that the things that are of most concern to religion are things that cannot be smelled, or touched, or tasted—such as freedom and rationality, good and evil, truth and falsehood, love and beauty. It is true that these will never register in the devices of experimentalists, or appear as quantities in the equations of theorists. But as we have seen, religion does make claims even about the physical world. One would be quite justified in calling these claims predictions.

One of these predictions is that the physical world cannot be "causally closed." One hundred years ago this prediction seemed well on the way to being falsified. Everything in the history of physics up to that time pointed toward a most rigid determinism. And yet that determinism did in fact give way in the face of new discoveries. Why shouldn't this be counted as a successful prediction?

Not only was classical physics deterministic, but it left no room for anything but matter. It is not simply that classical physics was solely about the behavior of matter—that, of course, is what physics is supposed to be about. It was that there was no way for anything else to enter the physicist's picture of reality. The formulation of classical physics was in terms of material systems, and the coordinates or "variables" that completely describe those systems, and the laws that completely govern those variables. To be brought within the discussion at all was to be brought within the realm of matter.

How could mind possibly have entered such a framework? The mind could only have been conceived of as yet another physical substance or phenomenon, describable by variables and governed by laws. This is the problem that bedeviled Descartes, who was both a physicist and philosopher. In his theory of physics matter could only be influenced by being pushed on by other matter that was touching it. How could the mind do anything, therefore, unless it too was a kind of matter that could touch things and push them around? Descartes, who rejected materialism, could not resolve this problem. The details of Newton's theory of physics were different from Descartes's, but it too portrayed a world of matter and motion and forces—a nuts-and-bolts world.

And yet, again, physics made a surprising turn. Suddenly the theories of physics could not even be formulated, it seemed, without reference to observers. Suddenly—as Heisenberg said—knowledge entered the picture, and—as Peierls said—"someone who knows" entered with it. Remarkably, this knower did not enter as just another physical system, with just another set of variables. The knower appears in the framework of the theory in a logically different way, on a different level, as it were. Is this not what the non-materialist view would have led one to expect? Is it not totally contrary to what the materialist view would have led one to expect? Is the materialist entitled to carry on as though his worldview were marching from triumph to triumph? Of course, no one knows what the future of science will bring. Perhaps quantum theory will itself be overturned. We can only talk about the implications of the science that we have.

From another direction, the direction of mathematics and logic, and at about the same time, the mechanical view of the human mind suffered at least as great a blow. The things that mathematicians understand can always be reduced to symbols and formulas and mechanical rules for manipulating them. But the very acts of "understanding" which are needed to create these formalisms cannot themselves be reduced to a formalism; they cannot be explained as mechanical processes of symbol manipulation, as mere algorithms. The whole mind of the mathematician cannot be reduced to mathematics. Mathematicians invent and use formalisms, but there is always something—the intellect—that remains outside the formalisms and judges truth, has insights, and applies the standards of rationality. Mathematical symbols can "mean" something, but there must be someone to whom they mean something. Arguments can be valid or invalid, but there must always be a mind to understand the arguments. To bring all the mental processes of a reasoning being within some finished mathematical description proves to be impossible. This seems to be the lesson of Gödel's Theorem.

How strikingly parallel this is to what is found in physics! The things that physicists know can always be reduced to equations and laws. But the very acts of "knowing" that give rise to these theories cannot be described by them. The observer cannot be absorbed into the systems he studies. There is always something that remains outside the system to "observe" it. There is knowledge, but there must be someone to know it. To bring all the processes of a knowing being within a closed physical description proves to be impossible, at least within the traditional framework of quantum theory.

Can this parallelism be entirely without significance? Is there not a lesson here? These are the most profound discoveries in mathematics and physics. Each deals with aspects of what it is to know; in one case to know through pure reason, and in the other to know through physical observation. In each case the totalistic dream of describing everything by a mathematical formalism or a law of physics runs into a contradiction: the contradictions pointed out by Lucas on the one hand and by von Neumann on the other.

The natural conclusions—as some people see them—of these great discoveries can be avoided. But in each case they can be avoided only by denying one's own status as someone who knows. In the Gödelian case, one can deny the consistency of one's own mind, and claim to be nothing more than an "inconsistent machine." In the quantum case, one can claim that what one knows by observation is not really the truth, but only that which is true in one branch of reality. The materialist seems to be forced to assert of himself not only that he is a machine, which for most people is absurd enough, but that he is really an infinite number of inconsistent machines dividing and subdividing into more and more realities as the universe unfolds.

Why should anyone prefer these alternatives to the straightforward belief that there is such a thing as an intellect? The reason is simple. In the eyes of many people, to accept the idea that the mind is something which cannot be reduced to mathematical description is to accept irrationalism. This comes out over and over again in the writings of materialists. Penrose, who in spite of his rejection of the "computational" view of the mind cannot bring himself to reject materialism, brands the idea that there is something immaterial about the mind as "the viewpoint of the mystic," and writes: "I reject mysticism in its negation of scientific criteria for the furtherance of knowledge."[1] Chalmers, who rejects materialism but cannot accept that the mind can have any actual effects in the physical world, defends his theories as showing that "to embrace dualism is not necessarily to embrace mystery."[2] In his view, to deny the possibility of a reductive explanation of the mind is "mysterianism."[3] Ernest Nagel and James R. Newman, who were among the first to argue that Gödel's Theorem implies a difference between the human mind and calculating machines, warned readers that "Gödel's proof should not be construed as . . . an excuse for mystery-mongering."[4]

What lies behind this terror of "mystery" is the idea that to understand something rationally is the same thing as to understand it through laws and equations and quantities. This identification, however, is far from being self-evidently justified. To some it came as a surprise that it is even justified in physics. Wigner, in a famous essay, wondered at what he called "the unreasonable effectiveness of mathematics" in understanding the physical world. He wrote, "The miracle of the appropriateness of the language of mathematics for the formulation of the laws of physics is a wonderful gift which we neither understand nor deserve."[5] Many other physicists have echoed these feelings.[6] Whatever the reason for it, experience has certainly shown the enormous power of mathematics in the realm of physics. But what gives us the right to expect that all of reality is re-ducible to such mathematical treatment? How often are the questions we ask in life answerable by equations? And yet, if the answers cannot be reduced to equa-tions, are they for that reason to be regarded as irrational? Is all of wisdom, all of morality, all of beauty, all of understanding a matter of numbers and laws?

There is a circularity about the materialist position that becomes obvious whenever its logic is carefully examined. The idea that everything may not be reducible to physics or mathematics is said to be mysticism, mysterianism, or mystery-mongering because it supposedly involves a rejection of rational expla-nation. That, in turn, follows from the supposition that all rational explanation must be explanation in terms of equations and quantities. This supposition is based on the fact that such quantitative explanations have been found to be suf-ficient in the realm of physics and on the assumption that what is true in physics must be true of all of reality. But what justifies that last assumption? Why, simply the idea that all of reality is nothing but physics!

So we come full circle: it is said that materialism is true because materialism is true, because it *must* be true. We saw the same circular reasoning applied to the origin of the human soul: human beings must be reducible to matter, it is said, because anything non-material about human beings could not have arisen by physical processes; and if it cannot have arisen physically, it cannot have arisen at all—certainly it cannot have been created by God. And this, finally, follows from the fact that only physical processes exist. In other words, materi-alism is true because materialism is true. It is certainly conceivable, if to many of us not credible, that materialism is true, but surely it is not irrational to ask for somewhat stronger arguments on its behalf.

Appendix A
God, Time, and Creation

The first sentence of the Bible is "In the beginning God created the heavens and the earth." (Gen. 1:1) This text has always been understood to teach both that the world was created by God and that it had a beginning in time. As explained in chapter 4, the first does not necessarily imply the second. The world could be created by God and nevertheless have always existed. The example used in chapter 4 to illustrate this idea was a lamp illuminating an object. Even if the lamp has always been illuminating the object, one can nevertheless say that the illumination of the object is caused by the lamp.

However, it is very hard for us to separate the idea of creation from the idea of beginnings. This is because our idea of creation comes from human examples. When a human artisan "creates" something, in the sense of fashioning it, the thing that he makes does begin to exist when he makes it. Of course, when a human being makes something, he is operating on a physical level. A singer produces the sound of the song through the motions of the vocal cords. A painter produces a picture by moving the brush with fingers, hands, and arms. Thus, human "creation" always involves a chain of physical cause and effect, with the body of the human creator being a part of that chain. Since physical events always precede their physical effects in time, by virtue of the way our universe is constructed,[1] a human creator must exist before the thing which he makes — temporally before.

In thinking about Creation, it is very hard for our minds to break free of the limitations of this human analogy. We tend, even if we know better, to imagine God as a physical being who uses physical processes to fashion the universe. Of course, that is absurd, and is not the traditional concept of God and Creation. Nevertheless, this kind of crude anthropomorphism underlies some of the objections that are occasionally raised against the notion of a creator. If God created the universe, it is said,[2] then God must have existed "before" the universe, in a temporal sense, and this is not possible. It is not possible *whether or not* the universe had a beginning in time. For if the universe had a beginning in time, then there was no such thing as "before" the universe, because time itself came into being along with the universe, as St. Augustine clearly understood. But on the

257

other hand, if the universe has existed for infinite time, then again it is impossible to talk about what existed "before" the universe.

Of course, the Jewish and Christian teaching is that God is not a physical entity. The anthropomorphic language used of God in many passages of the Bible has been understood since ancient times to be metaphorical. God did not fashion the universe by physical movements that initiated other physical movements.

GOD AS THE "FIRST MOVER" AND "FIRST CAUSE"

Some readers may suspect that the traditional notion of Creation was cruder than I am admitting. Perhaps they are thinking of the famous proof of St. Thomas Aquinas of the existence of a "First Mover."[3] This proof, which St. Thomas derived from Aristotle, starts with the principle that "nothing can move unless it is moved by something else." St. Thomas argued from this that there had to be a "First Mover," which he identified as God.

When people hear this argument they naturally imagine something like a game of billiards. A moving ball strikes another ball, which is at rest, and imparts some motion to it. This can happen in a series, with ball A imparting motion to ball B, which in turn imparts motion to ball C, and so on. What St. Thomas seemed to be saying is that, if you trace this series of collisions back in time, there has to have been a first moving ball. He seemed to be saying that God is like the cue ball that sets all the later balls in motion.

Thinking that this is what St. Thomas meant, many people dismiss his proof as simpleminded. They correctly note that from the point of view of modern physics there is nothing wrong with a hypothetical scenario in which objects (such as atoms) are perpetually colliding in an infinite series, without beginning and without end. It does not matter whether, in actual fact, they have been doing this; the point is that there is nothing physically or logically absurd about the idea that they have. Indeed, as we saw, before the discovery of the Big Bang, physicists such as Einstein were convinced that the motions of matter had indeed been going on forever.

However, this kind of sequence of moving objects is not at all what St. Thomas had in mind in the proof of a "First Mover."[4] If it had been, then, indeed, his conception of the Creator would have been a crude physical one. The reason that many people misunderstand St. Thomas is that he was using a technical philosophical terminology, in which certain words, in particular the word *move*, meant something quite different from what they mean in ordinary modern usage. For St. Thomas A "moves" B, if A is the cause of some effect in B. The effect in question does not have to have anything to do with movement through space.[5] St. Thomas was really thinking in very general terms about cause and effect.

St. Thomas distinguished between "simultaneously acting" causes and "non-simultaneously acting" causes. Non-simultaneously acting causes act one after another in time. For example, one can say that a person is caused, in some sense, by his parents; his parents are caused by his grandparents; and so on. Cause and effect, here, form a temporal chain or sequence. Now, St. Thomas saw nothing impossible in the idea of such a chain of non-simultaneous causes stretching back infinitely far into the past. In his *Summa Contra Gentiles* he considered the example of "a father being the cause of a son, and another person the cause of that father, and so on endlessly."[6] He concluded that there was nothing inherently absurd about this. He observed that, according to some philosophers, "in the sphere of non-simultaneously acting causes . . . it is not impossible to proceed to infinity."[7] That is, there is no reason that such a temporal sequence of cause and effect has to have a beginning. Therefore, he saw nothing clearly wrong, philosophically speaking, with the idea that the universe is infinitely old. Arguments that said that the universe had to have a beginning, as otherwise there would be an infinite chain of events without a first event, he curtly dismissed as having "no compelling force."

So, when St. Thomas said that there had to be a "First Mover," he did not mean some physical being, event, or motion that stood at the beginning of a sequence of physical movements. What, then, did he mean?

St. Thomas was thinking about "simultaneously acting causes."[8] He was not thinking of a time sequence at all. Consider the following example from physics. The mass of Earth gives rise to a gravitational field. Earth's mass can be said to be the cause of the gravitational field, but Earth's mass does not "happen before" the gravitational field. It is not that there was a time when Earth existed without a gravitational field, and then the gravitational field came into being. The gravitational field is in a real sense simultaneous with the mass which is its source.[9] Or, to use another kind of example, I may be envious of someone because I know that he is more gifted than I am. This need not imply that there was a time when I knew of his superiority and was not envious: I could have been envious since the first moment I knew of his superiority, and yet the knowledge is a cause of the envy. Perhaps this is easier to think about if one substitutes the word *explanation* for *cause*: A can be the explanation of B, without coming before B in time.

Now suppose that there is some set of simultaneously existing things, A, B, and C. Would it make sense to say that A explained B, B explained C, and C explained A? Most people would reject that as a form of "circular reasoning." If I say, "I am upset because she insults me; she insults me because I am fat; I am fat because I eat too much; and I eat too much because I am upset," then I am explaining in a circle. Instead of thinking of this as a circle, one can think of it as an infinite chain without beginning or end. That is, using A → B to mean "A is the cause (or explanation) of B," we have . . . A → B → C → A → B → C → A → B → C → A → B →. . . . Everything in this chain appears to have something before it "explaining" it, but really nothing is ultimately explained. St. Thomas argued that to have a real explanation, the explanation must start somewhere.

The same is true of mathematical proofs. A proof must start somewhere. One must start with some statements whose truth is unproven — that is, the axioms or first principles — and from them derive in a step-by-step manner the truth of other statements. A mathematical proof which went around in a circle would be inadmissible. A mathematical proof which went in an "infinite regress" would be equally inadmissible. An infinite regress is an infinite chain of reasoning that has no beginning. In an infinite regress, each statement can be shown to be true *if* the statements before it are true. But there are no statements at the beginning of the proof whose truth guarantees the truth of all the statements that follow.

When St. Thomas spoke of a First Mover, he was not speaking about a chain of events in time, he was speaking about an *explanatory* chain.[10]

Let us now re-examine the example, mentioned by St. Thomas, of an infinite series of fathers and sons. In that sequence of "non-simultaneously acting causes," each son's existence is caused, in a certain sense, by his father. *That* kind of chain of events, happening one after another in time, can certainly go back infinitely far. If it does, one might imagine that everything has been explained: the existence of each son has been "explained" by the existence of his father. However, that is false; not everything has been explained. Many further questions arise, such as, "What enables fathers to beget sons?" The answer to that question lies in biochemistry, and is based on the chemical properties of various atoms and molecules. But the biochemical explanation of reproduction will, itself, raise further questions, such as, "How are the chemical properties of atoms and molecules to be accounted for?" This, in turn, leads one to a deeper explanation. We therefore end up with another chain, but it is not a chain of events stretching back into the past. It is rather a chain that goes deeper and deeper, to more and more fundamental explanations. It is *this* kind of causal chain that St. Thomas had in mind, and which he maintained had to have a "first" term — but not first in the sense of "earliest" in time, rather "first," as he put it, "in the order of causes."[11]

GOD AS THE SOURCE OF BEING

Applying this to the universe, the point is that it does not matter whether it had a beginning in time or has an infinitely long history. In either case it must have some ultimate explanation. The contemporary philosopher Richard Swinburne states the point as follows:

> It would be an error to suppose that if the universe is infinitely old, and each state of the universe at each instant of time has a complete explanation in terms of a previous state of the universe and natural laws, [so that God is not

invoked], the existence of the universe throughout infinite time has [thereby been given] a complete explanation, or even a full explanation. It has not.[12]

As P. C. W. Davies said, in the passage quoted in chapter 4, we would still "be left with the mystery of why the universe has the nature it does, or why there is a universe at all."[13]

This last statement brings us to the deepest issue, the issue of existence. The question of "Creation" is not whether the universe had a definite point in time when it began, or whether it has lasted for an infinite time in the past. It is, rather, why there is a universe at all. Or, as the question is often put, "Why is there something, rather than nothing?"

One can compare different models of the universe to different kinds of geometrical lines. The Big Bang as described by Einstein's equations is like a line that starts at a definite point. The models where the universe is eternal (such as Einstein's eternal universe, the steady state universe, the bouncing universe, eternal inflation, and so on) are like lines that extend infinitely in both directions. The quantum scenarios where the universe has a finite age but no initial "boundary" are somewhat like lines of finite length which have no endpoints — like a circle. But the real issue is not the shape of the line. The issue is why there is any line at all. Where did it come from? Who drew it? This point is made succinctly by the cosmologist Don Page:

> God creates and sustains the entire universe rather than just the beginning. Whether or not the universe has a beginning has no relevance to the question of its creation, just as whether an artist's line has a beginning and an end, or instead forms a circle with no end, has no relevance to the question of its being drawn.[14]

CREATION AND TIME

Many people think of Creation as only being an "event" that happened at some time in the distant past. But this is not the whole story. The word *Creation* is used in two ways in theological writing. It is, indeed, often used to refer to the creation of the world *ex nihilo* at "the beginning of time." But while that is the more common meaning, it is not the most fundamental meaning of the term.

Creation in its deeper meaning refers to the divine act whereby God gives being to everything that exists. As St. Thomas expressed it, "God is to all things the cause of being."[15] Thus, every finite thing that exists, including every part of space and time, exists because God causes it to exist. It is the whole of the universe that is created in the sense that it is called into existence by God. The physicist Stephen Hawking famously asked about the laws of physics, "What is it that breathes fire into the equations and makes a universe for them to describe?" The theist answers, "God."

A traditional analogy compares God, the Author of the universe, to the human author of a book or a play. One can distinguish the *beginning* of the book, in the sense of its opening words or sentences, from the *origin* of the book. The entire book, not just the beginning, has its origin in the author's imagination. Every word of the text is part of the text by virtue of the author's decision that it should be. In the same way, the whole universe and all of its parts are equally created by God. This is the point that Don Page was making in the passage quoted earlier.

The events in a book have a certain sequence that constitutes the "timeline" of the book's plot. Within the plot of the book some events are causes of later events. Something that happens on page thirty causes something else to happen on page fifty. But the *author* causes the *book* at a different level altogether. On one level, Polonius dies because Hamlet stabs him, but, on another level, Hamlet stabs Polonius and Polonius dies because that is what Shakespeare wrote. The author causes every part of the book equally. Moreover, while events in the book, including its "beginning," have some definite location in the plot's timeline, the *creation* of the book has no such location: the author conceiving of the book's plot in his mind is not an event that takes place on a certain page *in* the book's plot. The author is external to his work, and his act of creation is not an episode within it. This leads to the conception, first clearly articulated by St. Augustine, that God is outside of the time of the universe. God is eternal, not in the sense of lasting for an infinite duration of time, but in the sense of being timeless.

GOD AS ETERNAL, OR "OUTSIDE OF TIME"

"Past" and "future" refer to the way that events in our universe's plot are causally related to each other within that plot; they pertain to the plot-time of the universe. But these relations do not apply to God's activity, which is on another level. Past and future do not apply to God, who, as St. Augustine expressed it, lives in "the sublimity of an ever-present eternity."[16] This perhaps can be better understood if one thinks of God as an infinite mind. As the great Jesuit philosopher Bernard J. F. Lonergan expressed it, God is the "unrestricted act of understanding" that perfectly grasps all that there is to understand, that is, all of reality.[17] But such a perfect and complete act of understanding cannot alter or vary. How could it, unless there was something it failed to grasp, or something that it grasped imperfectly? As truth is unchanging, so is the divine act that apprehends it in its fullness. In the traditional view, this infinite act of knowledge and understanding *is* God.

It therefore makes no more sense to talk of God changing than to talk of mathematical truths changing. One does not say "2 + 2 will be 4" or "2 + 2 was 4." Mathematical truths are tenseless, eternal. They just are. And that is how Jews and Christians have understood God to be. This is one meaning that theologians have seen in the name which God reveals to Moses in the Old Testament: "I AM WHO AM," or simply "I AM." ("Thus shalt thou say unto the children of Israel: 'I AM' hath sent me unto you." [Exod. 3:14]) God, the unchanging source of being, exists in a timeless present. St. Thomas called it the *"nunc stans,"* "the 'now' that stands still." This timelessness is also asserted by Christ in the New Testament, when he makes the astonishing statement about himself, "Before Abraham was, I AM" (John 8:58).

Thus, in the traditional understanding, Creation is an eternal act of God, in the sense of an act that is outside of time, even though the effects it produces, namely the things and events of this world, are within time, and are related one to another in a temporal sequence. The divine act of creation "precedes" the existence of the universe only in a causal sense, not in a temporal sense. It did not happen in a "time before" the universe came to be.

GOD AS THE "NECESSARY BEING"

The most common objection to the notion that God is the cause of the existence of the universe is that a cause must then be found for God. This is the old school-child's question: "Who created God?"[18] The traditional answer to this is that God's existence is "necessary." The analogy here is with contingent and necessary truths. An example of a contingent truth is that there is a sycamore tree in my front yard. This is true, but it did not have to be true; it just happens to be true. One may, therefore, legitimately ask how it came to be true, what caused the sycamore tree to be there. On the other hand, that 317 is a prime number is a necessary truth. It cannot have been otherwise. It makes no sense to ask how it came to be that way, or what caused it to be that way, at least not in the same sense that these questions can be asked about the sycamore's presence in my front yard. By analogy, it is argued that while some beings are "contingent beings," which could have existed or not, God is a "necessary being." And, like the necessary truths of mathematics, he is eternal and unchanging. For this reason, it is maintained that it makes no sense to talk about the cause of God. He is uncaused.

To sum up, one kind of argument for God's existence runs as follows: (1) Every contingent being or fact must have a cause or explanation. (2) Such causes cannot run in an infinite regress. (3) Consequently, there must be a First Cause. (4) Obviously, this First Cause must itself be uncaused. (5) This means that it must be a necessary rather than a contingent being (because of (1)). This necessary being is what is meant by God.

One "scientific" objection that is often made to this kind of argument is that quantum theory has shown that step 1 is invalid.[19] It is claimed that according to quantum theory certain contingent physical events are indeed uncaused. The classic example of an "uncaused" quantum event is the radioactive decay of an atomic nucleus. The laws of physics do not determine exactly when a particular radioactive nucleus will decay. It is a matter of probabilities. All that the laws of physics say is that if the nucleus has a half-life of, say, one hour then within one hour, there is a 50 percent chance for the nucleus to decay. If there were one thousand such nuclei, then after an hour approximately five hundred of them would have decayed. But the laws of physics would not say which ones had decayed or precisely when. Therefore, the fact that a particular nucleus decayed at a particular time is, in a certain sense, "uncaused," even though contingent.

However, this objection is not a cogent one. All that is really being said is that the laws of physics and the past state of the universe do not by themselves determine every event that will happen in the future. That is like saying that events in act I of *Hamlet* and the rules that govern the writing of plays do not by themselves exactly determine what happens in act II of *Hamlet*. Nevertheless, it would be foolish to argue on that basis that the particular words of *Hamlet*, or the play *Hamlet* as a whole, are uncaused, or that *Hamlet* needs no author, or that Shakespeare did not exist. To put it another way, all that quantum theory says (according to this interpretation of it) is that certain events do not have a completely determinative *physical* cause. That does not imply that these events have no cause whatsoever. That would only follow if one had already assumed that materialism is true and that all causes have to be physical causes.

Another common objection to the argument for a first cause is that all it shows is that *some* necessary being must exist, not that the necessary being must be God. In the words of Victor Stenger, author of *Not By Design*, "Later philosophers . . . have pointed out the error in Aquinas's logic: if a first cause, uncaused, is possible, why must it be God? The first cause, uncaused, could just as well be the universe itself."[20] That would, of course, be true only if the existence of the universe itself were necessary rather than contingent. Some have suggested that it might be. For example, the great eighteenth-century French physicist Jean d'Alembert wrote: "To someone who could grasp the universe from a unified standpoint, the entire creation would appear as a unique truth and necessity."[21]

The main problem with this idea is that it is patently absurd. The existence of the particular universe in which we live is plainly *not* a necessity. In this particular universe there is a sycamore tree in my front yard. It might just as well have been an apple tree. To say that this universe, in all its particularity, with all of its details, had necessarily to exist is not only absurd, it is also profoundly unscientific in spirit. It would mean that everything about the world could be deduced by pure thought without taking the trouble to do any experiments or make any observations. If the world with all its contents were necessarily as it is, then Columbus did not have to sail the ocean blue — he might have been able to deduce the existence of America and even to have mapped all its mountains and charted all its waterways without leaving his armchair.

God as "Eternal Reason"

One of the common misconceptions about belief in God is that it is based on there being an irrational element to reality. This is what some people think religious "mysteries" are all about. But quite the opposite is the case. Belief in God is based on the idea that reality is utterly rational. The theist believes that if something is part of reality it can be understood and known by a rational mind. There is, in other words, some act of knowing that can fully grasp it. The idea that all of reality is rational, intelligible, and knowable leads the theist to conclude that there is some act of knowing and understanding — infinite of course — that *completely* grasps all of reality.[22] It leads him to conclude, in fact, that to be a part of reality is precisely to be known by this supreme act, which is God. Such an all-embracing act of knowing and understanding must also know and understand what mind is, and what it means to "know" and to "understand." It must, indeed, perfectly know and understand *itself*. It is the supreme reality, which comprehends — in both senses of the word — everything that exists.

A religious "mystery," therefore, is not something that cannot be understood; it is something that cannot be fully understood *by us*. The theist does not believe that reality is impenetrable by reason. On the contrary, he believes that there is a Reason, a rational intellect, which completely penetrates it. Reason is that which grasps the connections among things, and therefore the infinite divine act of understanding is the act of reason that subsumes all acts of reason. In the words of St. Augustine, God himself is "eternal Reason."[23]

The Gospel of St. John begins with an account of Creation, which intentionally harks back to the opening words of Genesis. Its first words are well known: "In the beginning was the Word, and the Word was with God, and the Word was God. The same was with God in the beginning. By him all things were made." (John 1:1–3) The Greek term that is translated "Word" here is *Logos*. But *Logos* can also mean "Reason" as well as "Word." Thus the opening of St. John's gospel could be read: "In the beginning was Reason, and Reason was with God, and Reason was God." The *Logos* also refers to that "eternal Word" which God "spoke" in creating the universe (the term *speak* being used metaphorically here). The roots of the idea of the divine *Logos* go back to Jewish conceptions of the eternally pre-existing Torah, or Law, which was also identified by the rabbis with the divine Wisdom,[24] often personified in the Old Testament.[25] The divine Wisdom is portrayed in the Bible as being "spoken" by God and being "with God" at the Creation:

> With you is Wisdom, she who knows all your works, she who was present when you made the world . . . she knows and understands everything." (Wis. 9:9,11)

> I [Wisdom] came forth from the mouth of the Most High." (Sir. 24:3)

> But the One who knows Wisdom has grasped her with his intellect." (Bar. 3:32)

Thus, St. Augustine writes,

> In the beginning, O God, you made heaven and earth in your Word, . . . in your Wisdom, in your Truth. . . . "How great are your works, O Lord; you have made all things in wisdom!" (Ps. 104:24) That Wisdom *is* the beginning, and in that beginning you have made heaven and earth.[26]

What Augustine is saying here is that the "beginning" of which both Genesis and St. John's gospel speak refers not simply to a time, but also to the timeless *origin* of things, and he identifies that origin with the eternal Wisdom, or Word, or Reason of God, which is God himself.

The ancient Jewish sages had a similar understanding. In commenting on Genesis 1:1, they said, "And the word for 'beginning' refers only to the Torah, as scripture says, 'The Lord made me [Wisdom] as the beginning of his way.'"[27]

REASON AND FAITH

These reflections can help us understand something about faith and its relationship to reason. Even the atheist, precisely to the extent that he is rational, has a certain kind of faith. He asks questions about reality in the expectation that these questions will have answers and that these answers will make sense. Why does he believe this? It is not because he already has the answers and can therefore see that they exist and make sense — if he already had these answers he would not be seeking them. Yet he has the conviction that these rational answers exist. This is a faith. It is a faith that reality can be known through reason. It is a faith that those particular, limited acts of understanding through which he will grasp the answers to his questions are there waiting for him, so to speak, even if he does not succeed—even if no human being ever succeeds—in reaching them. It is what drives the scientist in his all-consuming quest. This faith, far from being opposed to reason, is a faith *in* reason.

The faith of the theist is of this kind. He has faith not only that there are some limited acts of understanding through which he will grasp the answers to his particular questions, but that there is a perfect and complete act of understanding which leaves no further questions to be asked. This complete act of understanding is God. For the believer, faith in God and faith in reason are profoundly linked. The Book of Genesis asserts that human beings are created "in the image of God." This has always been understood to refer primarily to the fact that human beings have rationality and freedom. Our reason, finite and limited though it is, is a reflection of the infinite divine Reason. Like God, therefore, we can grasp the world by our intellects, though unlike God, we can only do so partially. Both the rationality of the world and our capacity to understand it have the same ultimate source.

Appendix B
Attempts to Explain
the Beginning Scientifically

In chapter 7, I discussed some of the speculative scenarios in which time has no beginning and the Big Bang is merely the beginning of one part of the universe or one epoch in its history. Another line of physics speculation accepts the idea that time has a beginning, either the Big Bang that occurred some 15 billion years ago, or some earlier perhaps even bigger Bang, but seeks to give that beginning a scientific explanation. Many scientists are under the impression that such an explanation would render a divine creator superfluous. As I will explain later, this notion is based on a misunderstanding of the idea of Creation. However, let us put that issue aside for now and focus on the scientific ideas.

The Reasons to Look for a Theory of the Beginning

Theories of the beginning of the universe generally are formulated within the field called "quantum cosmology." There are several motivations for this work. At the most basic level, scientists seek to understand phenomena, and the Big Bang is a phenomenon. (In this appendix I will refer to the beginning of time as "the Big Bang," whether it is the Big Bang that happened about 15 billion years ago or some earlier event.) Scientists want to understand it in the sense of finding an adequate mathematical description of it in terms of the laws of physics. There is no reason why the Big Bang should not have such a description, and if such a description is possible it is an important goal of science to find it.

A second reason to study the Big Bang is that a correct scientific understanding of it is likely to give us a better understanding of what happened after the Big Bang. For example, an important unsolved problem in physics is to explain the so-called "arrow of time." Why is there such a profound asymmetry between past and future? The "microscopic" laws of physics that we so far know are (for the most part) "time symmetric" and do not explain time's arrow. The arrow of time is connected with the Second Law of Thermodynamics, which says that entropy increases as one goes from past to future. Many physicists think that the arrow of time is due to the fact that the universe had extremely low entropy at the time of the Big Bang, but why the initial entropy should have been so low remains somewhat mysterious. It is to be hoped that when the Big Bang is better understood we will have answers to such questions.

The Initial Singularity

Another motivation for research on the Big Bang is the desire to avoid what is called the "initial singularity." The word *singularity* has several meanings in mathematics, but it generally refers to a special point where something breaks down, often because some quantity becomes infinite. For example, in the equation $y = \frac{1}{x}$, the point $x = 0$ is singular because at that point $y = \frac{1}{0}$, which is infinite. When an expanding universe is described using Einstein's equations, it is found that there is a first moment of time, which it is convenient to call $t = 0$, where both the density of matter and the "curvature of space-time" are infinite. (In fact, the matter density and curvature both are proportional to $\frac{1}{t^2}$ near $t = 0$.) The point $t = 0$ is the so-called "initial singularity" of the universe.

Physicists do not like infinities in the solutions to their equations, or rather they tend not to believe them. The reason for this is that when infinities do show up in solving the equations that describe a physical system, it is almost always due to the fact that the system in question has been "idealized" in some way. For example, an ideal mathematical cone has infinite curvature at its tip, but the actual conical objects that are found in the real world are not perfect cones and do not have infinitely sharp points.

In physics one must often make what are called "simplifying assumptions." It is necessary to do so, because real physical systems are far too complex to study exactly. For example, if one wants to understand the motion of a billiard ball, it makes sense for many purposes to treat the ball as if it were a perfect sphere of infinite rigidity. To treat it exactly would mean worrying about every microscopic scratch on its surface, its chemical composition, and all sorts of other details. Indeed, it would ultimately mean keeping track of every one of its atoms, which would be impossible in practice.

So, scientists have to make approximations. Sometimes they go too far, like the physicist in the old joke scientists like to tell about the "spherical cow approximation." But a vital part of scientific training is learning how to make sensible and useful approximations. It frequently happens, because of some approximation or idealization of a system that has been made, that the calculation of a physical quantity gives an infinite result.

Suppose, for example, that one were to calculate the force experienced by two billiard balls when they collide, using the approximation that they are absolutely rigid spheres. One would find that the force, at the "instant" of collision, is infinite. Of course, that is not realistic. What happens when real billiard balls collide is that they get slightly deformed, stay in contact for some small but finite interval of time, and then spring apart. During that interval of contact the force is large but not infinite. The more rigid the balls are, the less they get deformed, the less time they remain in contact, and the greater the force they exert on each other while they are in contact. Real billiard balls, however, are not infinitely rigid, and the forces they undergo are never infinite.

Long experience has taught physicists that when they find infinities in their answers it is generally due to some invalid approximation, rather than some infinity in the real world. Therefore, it is quite natural for physicists to have the same suspicion about the "initial singularity" of the universe, and to try to find a more exact description of the Big Bang that gives finite answers.

Moreover, physicists already are aware of one improper idealization that they are making when they describe the Big Bang using Einstein's equations: they are leaving quantum theory out of account. Normally, this is justified when dealing with gravitation, because "quantum gravity" effects are ordinarily extremely small. In fact, no quantum gravity effect has ever been observed. However, theoretical arguments imply that when the density of matter and the curvature of space-time get large enough quantum gravity effects can no longer be neglected. Such conditions were certainly realized in the Big Bang. Most physicists who think about these issues expect that in the proper quantum theoretical description of the Big Bang the initial singularity will turn out to be "softened" or washed out. Whether this happens and how it happens are obviously important questions for study. Unfortunately, at the moment, no exact quantum version of Einstein's theory of gravity exists. Nevertheless, enough is understood about quantum theory and gravity to make some interesting speculations, and this is what quantum cosmologists do.

Initial Conditions

There are other considerations that motivate work in quantum cosmology. Some of them have to do with what are called "initial conditions." If one wants to calculate the behavior of a physical system, it is not enough to know what the laws of physics are. One also has to have sufficient information about what that system was doing at some particular time. For example, to calculate how a baseball is going to move after it leaves the bat, one has to know exactly where it was when it left the bat, how fast it was moving, and in what direction it was moving. That means knowing six numbers: three for where the ball is in three-dimensional space, one for its speed, and two for its direction. If one specifies the condition of a system at the beginning of a process, that is called specifying the initial conditions.

One can also specify the conditions at the end of a process, so-called "final conditions." For example, if one knows exactly how a home-run ball is moving as it goes over the fence, one can calculate how it was moving before it got there. Actually, there are many kinds of "boundary conditions" used in physics besides initial and final conditions. However, initial conditions are probably the most commonly used, because we know the past and do not know the future.

The same considerations apply to that enormous physical system we call the universe. To describe the universe completely, some appropriate boundary conditions must be specified. There is a problem, however, in specifying *initial* conditions for the universe: many physical quantities are infinite at $t = 0$, the initial singularity, and therefore cannot be specified at that point. One could always use other kinds of boundary conditions for the universe instead. One could, for example, specify the condition of the universe at some time after the beginning. But somehow that does not seem like the natural thing to do. One would like to be able to say how things started off and work forward from there. This is an additional reason why many physicists would like to get rid of the initial singularity.

However, some physicists would not be happy just finding a non-singular beginning for the universe where its initial conditions could be specified. They would prefer to avoid the necessity of specifying initial conditions for the universe altogether. The reasons for this are both aesthetic and philosophical. We have every reason to believe that the fundamental laws of physics are extremely beautiful mathematically, and in some sense even simple. (This was discussed in chapter 12.) However, as a rule, initial conditions are neither simple nor beautiful. They are just some set of numbers that do not have to have any rhyme or reason to them. In the baseball example that I used earlier, only six numbers were required to specify the initial conditions. But to specify initial conditions for the entire physical universe would presumably require a vast set of numbers—perhaps an infinite number of numbers!

This is the aesthetic motivation for trying to avoid initial conditions. The philosophical motivation is what might be called "theophobia." One can think of the initial conditions of the universe as the cause, in some sense, of all the later developments that take place. But who set those initial conditions? One obvious candidate is God, and indeed this is precisely the role that is assigned to God by "deists": he "wound up the universe like a watch" and let it go. Atheists would like the cosmic "watch" to be the self-winding kind.

Quantum Creation of the Universe from "Nothing"

A very interesting idea that has come out of quantum cosmology is that the creation of the universe from nothing can be explained as a "quantum fluctuation."

This is thought by some people (both atheists and theists) to be a scientific alternative to the idea of a creator. As we will explain later, it is not. But to appreciate the real significance of these ideas one must understand something of the physics involved. A full explanation would require the reader to take a course in quantum theory. Fortunately, this will not be necessary. A rough summary will suffice for our purposes.

What Quantum "Creation" Means

I will begin by describing the phenomenon of quantum creation in a more modest setting: the quantum creation of subatomic particles rather than the quantum creation of an entire "universe." In particular, I will describe the process called the "pair creation" of particles.

Every type of particle has an opposite type called its "anti-particle." By "opposite" I mean that certain properties of the anti-particle, such as its electric charge, are opposite to those of the particle. For example, an electron has an electric charge of -1, while its anti-particle, the "positron," has an electric charge of +1. On the other hand, in certain respects, a particle and its anti-particle are exactly the same, rather than opposite. For example, they have exactly the same mass. Some particles are their own anti-particles, like "photons"—particles of light.

If a particle and an anti-particle—say an electron and a positron—collide, they can "annihilate." The word *annihilate* literally means to reduce to "nothing" (in Latin, *nihil*), but that is not what actually happens when particles annihilate. Each particle has energy. Even an electron which is not moving has, as Einstein discovered, a "rest energy" equal to $m_e c^2$, where m_e is the mass of the electron. The same is true of a positron, which has the same mass as an electron, as we have already mentioned. Together, then, an electron and a positron have a total rest energy of $2m_e c^2$. They will have more energy than that if they are moving, for then in addition to their rest energy they will have energy of motion or "kinetic energy." All this energy has to go somewhere when the electron and positron "annihilate," because, as we know, energy is "conserved." What happens (most of the time) is that gamma rays are emitted—very energetic photons—so the electron and positron "annihilate" into a burst of light.

The same process can happen in reverse. Under suitable conditions, light energy can materialize, so to speak, into an electron and a positron. This process is called pair creation.

Pair creation can also be caused by other things, for example an electric field. Imagine a volume of space permeated by an electric field, but otherwise completely empty. In that situation there is a certain probability that suddenly an electron-positron pair will pop out of empty space. The pair creation of particles by an electric field is an inherently quantum phenomenon. Such a thing cannot happen in classical (i.e., pre-quantum) physics. Let us see why.

One might imagine the pair creation happening in the following steps. (1) There is just empty space filled with an electric field, but no particles. (2) At some instant of time, the electron and positron suddenly appear at the same point of space. (3) The electron and positron move away from each other, pulled apart by the electric field (since they have opposite charges the electric field pulls them in opposite directions). As the electric field pulls them apart, they move faster and faster, becoming more and more energetic. The kinetic energy they thus obtain comes from the electric field.

The trouble with this scenario lies in step 2. The pair of particles suddenly appearing out of empty space would mean that the amount of energy jumps by $2m_ec^2$, as Einstein tells us. But the amount of energy cannot change, since energy is conserved. That is why, classically, an electric field cannot cause pair creation. So how does it happen in quantum theory? It happens by a process called quantum tunneling.

Quantum tunneling can be thought about in many ways. One way is the following. When the pair of particles suddenly appears at zero distance from each other they have rest energy equal to $2m_ec^2$. Let us suppose that to compensate this they have kinetic energy, i.e., energy of motion, that is *minus $2m_ec^2$, a negative* amount! Then their total energy would be zero, and it would cost no energy to produce them. Therefore, the principle that energy must always be conserved would not prevent them from being produced.

But how can the particles have negative kinetic energy? Classically, they cannot. Classically, the smallest amount of kinetic energy that something can have is zero—when it is not moving. When it is moving, it has positive kinetic energy. In fact, Newton tells us that the kinetic energy of a body is proportional to the square of its velocity; and the square of a number is never negative. Quantum theory, however, allows particles to have negative kinetic energy, even though only temporarily.

What happens in the pair creation of an electron and positron by an electric field can be thought of in the following way. The pair of particles is created with total energy zero: positive rest energy $2m_ec^2$ and negative kinetic energy $-2m_ec^2$. The pair of particles is pulled apart by the electric field, gaining in the process more and more kinetic energy, while the energy in the electric field is correspondingly reduced. At a certain point the kinetic energy of the particles has increased to the point that it becomes positive, as classical physics says it should be. Thereafter, the behavior of the particles can be described fairly well using classical physics.

One sees, then, that the actual process of "creating" the particles and the first stage of their existence (where they have negative kinetic energy) must be described using the strange rules of quantum theory. Classical concepts become inadequate. One might say that the conception and embryonic stage of the particles' life is only describable quantum-theoretically, whereas the post-partum period is describable fairly well classically. One of the classical concepts that breaks down during the embryonic phase of the particles' existence is time.

We saw that during their embryonic phase the particles can be thought of as having negative kinetic energy, even though, according to Newton, kinetic energy is proportional to the square of velocity. But if the square of the velocity is negative, then the velocity itself must be an "imaginary number." (Recall from chapter 12 that the imaginary number i is defined to be the square root of -1.) Since velocity is distance divided by time, an imaginary velocity means that the time interval is also an imaginary quantity. Thus, during the process of quantum creation of particles by "tunneling," time itself must be thought of using radically different concepts. This will be a very important point when we come to discuss the idea of quantum creation of the universe.

One thing that should be emphasized is that one cannot predict where and when such a pair creation event will happen. Because it is a quantum process it can only be predicted probabilistically. Quantum theory tells us that fields and particles are always undergoing so-called "quantum fluctuations." When these fluctuations get large enough, they can produce, say, a pair of particles. Another thing that should be emphasized is that this phenomenon of pair creation of particles by electric fields is not a matter of mere hypothesis or speculation. It has been observed in the laboratory, and physicists know how to calculate the probability of its happening, because they have an extremely well-tested theory that describes the behavior of electrons, positrons, and electric fields called quantum electrodynamics.

Quantum Creation of Universes

The idea of quantum creation of universes has been proposed in analogy to the quantum creation of particles. To grasp this idea, we first must understand what is meant by a "universe" here. A universe is a self-contained space having some non-zero volume and possibly filled with matter of some sort. I say "a" universe, because one can imagine a plurality of such universes, having different volumes, and shapes, and matter content. In chapter 7 we used the analogy of the surface of a balloon for our universe. We said that if the universe is expanding, it is like the surface of the balloon expanding, as it does when the balloon is inflated. Well, there is no reason one cannot imagine several different balloons, each balloon representing a different universe.

Just as we saw that a pair of particles can suddenly appear out of nothing (or rather out of an electric field) by a "quantum fluctuation," so one can imagine a tiny "balloon" suddenly popping into existence out of thin air by a quantum fluctuation. Of course, we do not mean "out of thin air" literally. The little universe actually would pop into existence out of . . . out of what? Out of nothing, really. Not even out of empty space.

Remember, space is something that is part of a universe. It is the surface of the balloon that represents space. Recall from chapter 7 that we are not to imagine any inside or outside of the balloon. The only space that there is is the rubber sheet of the balloon. We are to wipe away from our imaginations any thought of an inside or outside filled with air or anything else.

So we can imagine a situation where there are seven universes, and an eighth universe suddenly pops into existence by a "quantum fluctuation." One way that this can happen is that a small region of one of the universes (or balloons) gets "pinched off" to form a distinct universe. (The surfaces of the balloons quantum fluctuate, and one of these quantum fluctuations might happen to be large enough to pinch off.) Or it can just pop into existence apart from any of the already existing universes. One can also imagine a situation where there are no universes, and one universe suddenly pops into existence by a quantum fluctuation.

If this is the way things happened, then (as was the case with the pair creation of particles) the conception and embryonic stage of the universe we live in would have to be described using quantum concepts. This means that one could no longer take seriously the "initial singularity" that was a feature of the classical description of the Big Bang. In fact, our very ideas about the nature of time are likely to break down when applied to the beginning of the universe, for reasons already explained, so that even to talk about "an initial time" $t = 0$ is probably meaningless. But what concepts will be needed to speak correctly about what happened at the time of the Big Bang, no one yet knows.

The idea that our universe began as a quantum fluctuation is a very beautiful and interesting one, and quite possibly correct. However, unlike the quantum creation of particles, the quantum creation of universes remains at present just a speculative idea. We do have a theory—Einstein's theory of gravity (the General Theory of Relativity)—that describes the way spaces or universes behave, and it is very well tested. But people do not understand sufficiently how to calculate quantum effects in that theory. Therefore, no one really knows how to give an adequate mathematical description of such a process as the quantum creation of a universe. It is hoped that someday, superstring theory will tell us how to do this. Until that day comes, we cannot say whether the idea that our universe began by a quantum fluctuation really makes sense, let alone is correct.

WHEN IS "NOTHING" REALLY SOMETHING?

In quantum theory—as in the classical physics that came before—one is always interested in the behavior of some "system." The system can be as simple as a pendulum, or as complex as the whole universe. The system for a "condensed matter physicist" might be a piece of solid matter or a volume of liquid. For a nuclear physicist it might be the nucleus of an atom, or several nuclei interacting with each other. For an astrophysicist it might be a star.

In quantum theory every system no matter what it might be has a set of possible "states." The laws of physics, in their quantum-theoretical form, tell one how to calculate the probability that a system, which initially was in one state, will later be in some other state. Thus every state has definite characteristics and is related to the other possible states of the system by definite laws.

For example, in the case of the pair creation of an electron and a positron, one starts with "the system" in a state that has an electric field but no particles, and ends up with the system in a state that has an electron, a positron, and a somewhat different electric field. Note: the characteristics of the state have changed, but they are states of *the same system*.

How does one compute the likelihood that there will be a transition within a certain period of time from the former state to the latter? To perform such a calculation, one has to have a theory that describes in a precise mathematical way the interaction of electrons and positrons with electric fields. There exists such a theory, as was noted earlier, and it is called quantum electrodynamics, or QED for short. Within the framework of that theory, electrons, positrons, and electric fields act in certain prescribed ways. They do not act in an arbitrary manner. For example, in QED it is not allowed for a state with no particles to make a transition to a state with only electrons, or only positrons; the electrons and positrons must be paired. Moreover, there is greater probability for the state with no particles to make a transition to a state with one electron-positron pair than to make a transition to a state with two such pairs; and the relative probabilities can be precisely computed.

That is, the states of a system have very definite properties and relationships with other states of the same system, and these properties and relationships are governed by definite mathematical rules. It is important not to confuse a state of the system with the system itself. The system does not consist of, say, an electric field with an electron-positron pair; that is just one state of the system. The system itself is something more encompassing.

Now let us turn to "quantum creation of the universe from nothing." What one would be calculating there, in essence, is some sort of probability that a transition would occur from a "state" with "no universe" to a "state" with "one universe." (Or one might calculate the transition probability from a state with seven universes to one with nine universes.) No one really knows how to do such calculations yet, but let us not worry about that. Let us suppose that someday they will, and that the whole scenario will turn out to be correct. The key question for us is in what sense such a scenario can claim to explain the creation of the universe from "nothing."

In a sense it is correct to say, in the quantum creation of the universe scenario, that there was "nothing" before the universe appeared. There were no particles, and not even any space for particles to be in. Indeed, there was not even any time, so that using words like *was* and *were* and *before* is an abuse of language. Of course, all this was equally true in the classical (i.e., non-quantum) description of the Big Bang, with its initial singularity, that we are trying to go beyond. In both the classical and the quantum picture, there is a universe of finite age, before the beginning of which there was nothing, not even a "before." The more correct quantum description of the Big Bang really does not change anything fundamental in this regard.

Nevertheless, the idea of the universe appearing by means of a quantum fluctuation appears somehow to give a scientific explanation of how "nothing" can turn into something in a natural way. But does it really?

Let us look at the "no-universe state" that precedes (if we may use that word) the appearance of the universe. In a rather obvious sense the no-universe state is not really "nothing." It is a state! It is a specific "state" of a specific, complicated quantum "system." One can talk about a "no-universe state" only within the framework of a definite quantum system governed by definite laws. Within that framework, the no-universe state is one state among many, and has certain definite relationships to the other states. All those states are states of one system, and the equations used to calculate the transitions among those states follow from the laws governing that system, and in effect define the characteristics of that system.

One could imagine a *different* quantum system where the "universes" in question had not three but, say, seven dimensions, and had not the familiar kinds of matter, like electrons, but different kinds. Such a system would have *different* possible states. One of those states might be a state with "one universe," but a "universe" in this case would mean something quite different from what it meant before: it would mean something seven-dimensional filled with strange stuff. There would also be a "no-universe state" of this system. But this would be something quite different from the no-universe state of the system we contemplated before. It would be part of a completely different mathematical structure. For example, *this* no-universe state could make transitions to states with one or more seven-dimensional universes, but not to states with three-dimensional universes. The no-universe state we were discussing previously could only make transitions to states containing one or more three-dimensional universes. Now, clearly, if one can talk about different *kinds* of no-universe states—as I have just shown is possible—one is clearly not talking about "nothing."

In other words, there is a subtle equivocation going on about the word *nothing*. An analogy might help here. There is a difference between my having a bank account with no money in it, and having no bank account at all. The difference is not a spendable one. I may be broke in either case. But there is a difference nonetheless. For me to have a bank account—even one which has a zero or negative balance—presupposes several things: the existence of a bank, a system of money, an agreement between the bank and me that governs how money may be deposited or withdrawn (i.e., rules governing transitions between one state of my bank account and another), and so on. A complex system is in place. That system has many states. One state corresponds to there being one hundred dollars in my account, while another corresponds to there being no money in my account. That "no-money state," in certain practical terms, but not all, may be a lot like a situation where I have no bank account at that bank, or where there is no bank or banking system at all. But it is not the same situation.

WHAT "CREATION FROM NOTHING" REALLY MEANS

The idea that quantum theories of the Big Bang are competing against God as Creator is based on some crude misunderstandings. The real question that monotheism is attempting to deal with is a much more basic question than whether the Big Bang can be described by a mathematically consistent set of laws. It is rather why anything exists at all. Why are there any physical systems, any "states," any laws, any anything?

Just having a mathematical framework which describes a universe with a beginning, whether the framework is classical or quantum, or whether the beginning is smooth or mathematically singular, does not explain why that mathematical framework describes anything real. I can tell a story which gives a perfectly consistent and intelligible account of my becoming rich and explaining how I did it, but that does not imply that the story describes any real state of affairs. Such a story told about Bill Gates describes a reality; told about me it is a fantasy. One may contemplate some set of mathematical equations that purports to describe the beginnings of a universe. But is that a description of real events or of merely hypothetical events?

Physics will someday, one hopes, describe what happened at the Big Bang, but it will not tell us why it really happened rather than being just some mathematically self-consistent but never realized scenario.

The crucial question was lucidly posed by Stephen Hawking, who pointed out that a theory of physics is "just a set of rules and equations," and then went on to ask, "What is it that breathes fire into the equations and makes a universe for them to describe? The usual approach of science of constructing a mathematical model cannot answer the question of why there should be a universe for the model to describe."[1]

Appendix C
Gödel's Theorem

Gödel's Theorem is regarded as one of the great intellectual milestones, comparable in the depth of its philosophical implications to the discoveries in physics of Newton, or Einstein, or Heisenberg. I will try to give some idea of how Gödel proved it, but before doing so it may be helpful or at least interesting to sketch a little of the historical background of Gödel's achievement.

The first thing to appreciate is that mathematics is not just a collection of facts. Rather, what makes mathematics so interesting—at least to some people—is the fact that one can start with a few obvious and seemingly trivial statements and deduce from them a large number of things that are not at all obvious or trivial. The classic example of this was Euclid's treatise on geometry, written about twenty-three centuries ago, in which he based all of plane geometry with its many beautiful theorems about circles, and triangles, and other shapes on only ten fundamental axioms. For many centuries Euclid's achievement stood as the chief example of the power of pure reason, and as a model for mathematicians to emulate.

Later, mathematicians tried to find systems of axioms from which other branches of mathematics could be built up. Gottlob Frege wrote a book at the end of the nineteenth century called *The Foundations of Arithmetic,* in which he tried to construct an axiomatic system for arithmetic and algebra. Sadly for him, as he was getting ready to publish his *magnum opus,* Frege got a letter from Bertrand Russell pointing out a fatal inconsistency in a large part of his system. This inconsistency came to be called Russell's paradox.

Frege's difficulty arose from the carefree way in which he had used the notion of "set." (Actually, Frege used the notion of "concept" rather than "set." However, for the present discussion the difference is not important.) A set is just a collection of objects, which are called "members" of the set. One can have sets of anything: sets of apples, sets of oranges, or sets of numbers, for example. One

279

can even have sets of sets! In fact, modern mathematics could not get along without the idea of sets that contain other sets as members. This, along with the fact that he had used sets with an infinite number of members (which modern mathematics also cannot do without), was what got Frege into hot water.

Russell pointed out that if sets can contain sets as members, without any restrictions, then one could have the curious situation where a set contained *itself* as a member. Some sets would contain themselves as members, and others would not. Let us call a set which does *not* contain itself as a member a "normal set." Then one can think about the set—which in honor of Russell we will call **R**—that is defined to be the set whose membership consists of all normal sets. That is to say, **R** contains as members all those sets, and only those sets, that do not contain themselves as members. (Notice, by the way, that the set **R** has an infinite number of members.)

The deadly question—for Frege—was "does **R** contain itself?" There is no consistent answer. If **R** does *not* contain itself, then it is a normal set—by the definition of what a normal set is—and therefore it qualifies for membership in **R**, because **R** contains all normal sets. In other words, if **R** does *not* contain itself then it *does* contain itself. On the other hand, if **R** *does* contain itself, then it is not "normal" and therefore it does not qualify for membership in **R**. So if **R** *does* contain itself, then it does *not* contain itself.

What a logical predicament! No wonder that Frege was upset. (He was so upset that he uttered the famous lament, "Arithmetic is tottering"! Of course, as Frege understood, arithmetic wasn't tottering, only his attempt to axiomatize it was.)

The same paradox can be stated without using the word *set*. There is, for example, the "barber paradox": There is a barber who shaves all the men in town and only those men who do not shave themselves. Does the barber shave himself?

Russell's paradox is an example of the contradictions that can arise as a result of what is called "self-reference." A set containing itself is self-referential: if one tried to define it by listing all of its members, one would have to list the set itself in its own definition. This is like a word appearing in its own definition in the dictionary. It ends up chasing its tail. Not all self-reference is necessarily illegitimate. For example, "concept" is a concept. "Abstract" is abstract. "Meaningful" is meaningful. But there are other kinds of self-reference that are clearly improper. A well-known example is the statement, "This statement is false." If it is true then it is false, and if it is false then it is true.

Because of paradoxes like Russell's, mathematicians became aware that mathematical ideas and ways of reasoning that seemed at first glance inoffensive and clear could contain hidden traps and inconsistencies, especially if infinite sets were involved. And therefore a movement for greater carefulness and logical rigor grew up among mathematicians. One of the leaders of this movement was David Hilbert. Hilbert's proposal for achieving greater rigor was to completely "formalize" mathematics.

I have already explained, in the first section of chapter 22, what formalizing a branch of mathematics entails. Mathematical statements are written in a definite system of symbols, and one reduces all of the kinds of mathematical reasoning one is interested in doing to a precise set of rules for manipulating those symbols. One of the advantages of this way of doing things is that it leaves no room for confusion, or conceptual errors, or subtle ambiguities, or vagueness. In fact, as long as the rules of one's "formal system" are consistent and one adheres to them, there is no room for error at all. In this way absolute rigor can be achieved in mathematics.

A vital question, of course, is whether the rules of the formal system are actually consistent. Consistency, as I said in the second section of chapter 22, means simply that it is impossible to derive both a proposition, P, and its contrary, ¬P, from the axioms. (Henceforth I will use capital letters to refer to propositions, and the symbol ¬ to mean "not." In other words the negation of the proposition P is the proposition ¬P.) Hilbert hoped that if a branch of mathematics were completely formalized, methods would be developed that would allow mathematicians to check that its rules and axioms were completely self-consistent. In that way, mistakes like Frege's could be avoided, and all branches of mathematics could be built on solid foundations.

Suppose, for example, one wanted to check that the rules of arithmetic were self-consistent. One could formalize arithmetical reasoning, and then check that the rules of formalized arithmetic never allowed one to derive a contradiction. Of course, this would lead to a further question: were the methods of reasoning that one used to prove that arithmetic was consistent, themselves consistent? One could try to formalize *those* rules, of course, and check their consistency. But then the same question would arise again. Were the methods of reasoning one used to check the consistency of *those* rules, themselves consistent? How can one avoid going on like this forever in an endless regress?

The only way is to prove that the formal system one is interested in is consistent *using only methods of reasoning that are available within that formal system*. Then one will have both proven the system's consistency and at the same time shown that the steps of reasoning used to do this were themselves consistent. All doubts about consistency would then be laid to rest forever. This is basically the problem that Hilbert set for the world of mathematics.

Formalizing mathematics may sound like a dry and unexciting thing, but it brings other benefits to the mathematician besides logical rigor. The theorems that the mathematician proves also gain in generality. We saw an example of this in the first section of chapter 22, where we proved a little theorem about addition: $a + b + c = c + b + a$. While ostensibly this was a theorem about addition, we saw that it remained true if the symbols are reinterpreted to mean other things. For instance, if + is interpreted to be the sign for multiplication the proof still goes through, so that it also yields a theorem about multiplication. When mathematics is formalized, the theorems that are proven can often be applied in a wide variety of contexts. The same theorem, for instance, might be interpreted as saying something about lines and points, for example, or about numbers.

However, what is gained in rigor and power by formalization is lost in meaning. For, even though the strings of symbols of a formal system can be given various interpretations, it is not really necessary that they be given any interpretation at all. In fact, one can easily invent formal systems without any interpretation in mind. In some fundamental sense, mathematics that has been formalized has been drained of meaning. To make this clearer, imagine a person who has "learned Latin" in the sense that he knows which words are nouns, verbs, adverbs, and so on; knows how to conjugate the verbs and decline the nouns and adjectives; knows proper spelling; and can put together sentences that satisfy all the rules of grammar and syntax—and yet does not know what any of the words mean. This may sound crazy, but, in fact, computer programs that translate from one language to another are in precisely this situation.

In mathematics conceived of in this purely "formalistic" way, mathematical statements do not "mean" anything. Nor can one say that a mathematical statement is "true" or "false" in the same way that statements about the real world are true and false. All that really matters is whether a proposition—a string of symbols—can be obtained from the axioms by use of the rules. It becomes a sort of game, like chess. In chess a "legal" position is one that can be obtained from the starting position by making moves that are in accordance with the rules. An "illegal" position is one that cannot. No one would regard chess positions as "true" and "false" in any ordinary sense.

Those who take this stance on questions of meaning and truth in mathematics are called "formalists." Such a stance is summed up in Bertrand Russell's famous remark that "pure mathematics is the subject in which we do not know what we are talking about, or whether what we are saying is true." Only a minority of mathematicians and philosophers of mathematics take this extreme view of things.[1] Indeed, Gödel's Theorem is generally agreed to have created difficulties for the formalist viewpoint.

In any event, it was in answer to Hilbert's challenge that Gödel started the investigation that culminated in his great theorem. As we saw, what he showed was that in any consistent formal system that contains at least basic logic and arithmetic there are "undecidable" propositions: propositions which cannot be proven or disproven using only the rules of inference available in that system. And among those undecidable propositions is one that, in effect, asserts that the system itself is consistent. In other words, a final mathematical demonstration of the consistency of any formal system that contains at least basic logic and arithmetic is forever out of reach.

Gödel in essence showed the inherent limitations of formal systems. If they are consistent, then they are unable to prove their own consistency. And if one has a set of axioms that is consistent, then it is necessarily "incomplete" because it leaves many propositions "undecidable." Gödel's Theorem was the death of Hilbert's program of completely formalizing mathematics. No set of axioms, no formal system of rules, will ever be enough to capture all of mathematical truth.

How Gödel Proved His Theorem

It is easy to get a rough idea of how Gödel proved what he did, but the actual proof is mind-twistingly subtle. I will give the rough idea first, and then sketch out (in an only slightly less rough fashion) how the proof was actually carried out. Those who would like to know more about the proof of Gödel's Theorem should turn to the excellent popular account written in 1958 by Ernest Nagel and James R. Newman entitled *Gödel's Proof*.[2] The explanation here is basically a re-presentation of the account given in Nagel and Newman's book.

A Rough Idea of the Proof

Let us suppose we have a formal system, which we can call **F**. We will assume that this system is "sound." This is a stronger assumption than mere consistency. A system is sound if all the statements provable in the system are true, and not merely consistent with each other. (As G. K. Chesterton noted, a lunatic can have a perfectly self-consistent but false set of beliefs.) We are assuming that **F** is sound because it is easier to make the argument with this stronger assumption. Later we will make do with the weaker assumption of consistency.

Let us suppose that in the formal system **F** there is a statement which says, "This statement is not provable in **F**." ("Provable in **F**" means provable using the rules of inference of **F**.) In other words, we have a statement, which we will call G, that says "G is not provable in **F**." Notice that G says something about itself—that is, it is self-referential. I will return to this point later.

Now let us assume, hypothetically, that G is provable in **F**. That means one can go ahead and prove G in **F**. What would that imply? Well, first of all, because **F** is sound, anything that is proved within **F** must be true. Since G has been proved in **F**, then G must be true. But what G asserts is precisely that G is not provable in **F**. So it must be really *true* that G is not provable in **F**. However, that contradicts the assumption with which we started at the beginning of this paragraph. We have landed in a contradiction. Since our starting assumption led to a contradiction, it must have been false. That is to say, G must *not* be provable in F after all.

Thus, I have given a "proof by contradiction" which demonstrates the fact that G is not provable in F. But I have done more. Because the thing that I have proven (namely, that G is not provable in F) is precisely the thing that G itself asserts, I have succeeded in proving that G is *true*. So, in one fell swoop I have shown both that G is true and that it cannot be proven within the formal system F. This result is all one really needs for the purposes of understanding the Lucas-Penrose argument (see chapter 22, page 213), but we can go further. Let us do so.

Now suppose hypothetically, contrary to what we assumed hypothetically before, that we can *disprove* G within F. That is, suppose we can prove the contrary statement, ¬G, using the rules of F. Then ¬G must be true—because F is sound. But what ¬G says is precisely that G *is* provable in F (the contrary of what G said). So we have the situation that G is provable in F (just demonstrated), and that ¬G is provable in F (assumed at the beginning of this paragraph). But since F is sound, that means that both G and ¬G are true, which is impossible, since a thing and its contrary cannot both be true. So again we reach a contradiction. Thus, the assumption made at the beginning of this paragraph was false. In other words, G cannot be disproved in F.

Since we showed (two paragraphs ago) that G is not provable in F, and have now shown that it is also not disprovable in F, we have shown that G is "undecidable" in the formal system F. That means the system of axioms of F is "incomplete." But the most interesting fact is that, though G is undecidable in F, we have nevertheless been able to show that G is true.

There are two things wrong with this account of Gödel's proof. The first is that we used the very strong assumption that F is sound, whereas Gödel (for most of his results) was able to get by with assuming only that F is consistent. But more seriously, we have used what appear to be "self-referential" statements; and we know that such statements are of doubtful propriety. The reader will recall the statement, "This statement is false." That kind of statement, which refers to itself, is known to lead to logical trouble. Are we any better off using statements like "This statement is unprovable in F"?

The answer is that statements like "This statement is unprovable in F" can be made logically acceptable if one is very careful about what one is doing. I will now attempt to give some idea of what being "very careful" means.

A Somewhat Less Rough Idea

The reason why the statement "This statement is unprovable in F" is not acceptable as it stands is that it confuses statements of two types. Suppose the formal system F is arithmetic. Then statements in F are statements about numbers and their relationships to each other. Examples of such statements are "2 + 3 = 5" and "if $x < y$, then $y > x$." On the other hand, consider the statement "2 + 3 = 5 is provable in **F**." This is not a statement about numbers, but a statement *about* a statement about numbers. Statements about numbers and their properties we can call "mathematical statements," while statements about mathematical statements we can call "meta-mathematical statements." They are different kinds of statements. Strictly speaking the same statement cannot be a mathematical statement and a meta-mathematical statement.

We supposed that we had a statement, which said, "G is unprovable in **F**." If G is a mathematical statement, then "G is unprovable in **F**" is a *meta-mathematical* statement. So the statement G and the statement "G is unprovable in **F**" are different kinds of statements. Therefore, one cannot have the situation assumed earlier where G *is* the statement "G is unprovable in **F**." So our "rough idea" of Gödel's proof was a bit of a cheat.

In view of this, let us modify the earlier argument as follows. Let G be a statement about numbers; that is, a mathematical statement. And let K be the meta-mathematical statement, "G is unprovable in **F**." The problem just discussed would arise if we said that the mathematical statement G and the meta-mathematical statement K were one and the same statement; so we will not do that. Suppose, instead, that we somehow yoke together the two statements G and K so that they stand or fall together. That is, if one is true then so is the other, and if one is false then so is the other. How this yoking is to be done we will see later. With such a yoking, we can carry through essentially the same proof as before, as we shall now see, but without having committed the sin of self-reference.

The first step is to suppose that G is provable in the system **F**. That means we can go ahead and prove G in **F**. Then because we assume **F** is sound, it follows that G is true. But since G and K "stand or fall together," it follows that K is true also. But K is the meta-mathematical statement "G is unprovable in **F**," and so, since K is true, it must actually be the case that G *is* unprovable in the system **F**. However, this directly contradicts the assumption with which we started at the beginning of this paragraph, namely that G is provable in **F**. That starting assumption has led us to a contradiction, and therefore must be false. Thus G is actually unprovable in **F**, which is to say that K is true. But as G and K stand or fall together, G is also true. We have thus shown both that G is unprovable in **F** and that it is nevertheless true. This is the result we want, and we have achieved it without using self-referential statements.

How to Do the Yoking

All of this was accomplished because of the yoking together of G and K. The question, then, is how to do the "yoking"? To do this Gödel used a concept very common in mathematics called "mapping." Two things that are not the same can be "mapped onto" each other if they have the same or similar structures. For example, if I have ten circles of unequal areas and ten stones of unequal weight, I can map statements about the relative areas of the circles onto statements about the relative weights of the stones. Thus the statement, "If circle A is larger than circle B, and circle B is larger than circle C, then circle A is larger than circle C" can be mapped onto the statement, "If stone A is heavier than stone B, and stone B is heavier than stone C, then stone A is heavier than stone C."

What mapping allows us to do is relate statements that are about different things, but which really, in a sense, say the same thing and therefore stand or fall together. What Gödel had to do, then, was to map mathematical statements and meta-mathematical statements into each other. In particular, since he was talking about arithmetic, he had to create a mapping between statements about *numbers* and statements about *statements about numbers*. That means he had to create a mapping between numbers and statements about numbers.

To get a crude idea of how this might be done, let us first invent our own very simple, and very silly, formal system. Let there be in it just three symbols: A, B, and C. The grammatical rules will also be simple: any string of A's, B's, and C's in any order will be a proper "statement" in the system. So, for example, "ACBBAC" is a proper statement. The rules of inference will be somewhat peculiar: one statement can be derived from another if it has the same "numerical weight." The numerical weight of a statement is defined to be the sum of the weights of its letters, where A, B, and C have the weights 1, 2, and 3, respectively. So the numerical weight of the statement "ACBBAC" is $1 + 3 + 2 + 2 + 1 + 3 = 12$. Finally, I choose the only axiom of the system to be the statement "ABC." (Note that the axiom has a numerical weight of $1 + 2 + 3 = 6$.) Let us call this formal system "alphabetics."

It is easy to see that, since the one and only axiom has the numerical weight 6, and since the rule of inference allows a statement to be derived from another if and only if it has the same numerical weight, then all statements of weight 6 can be derived from the axiom and are therefore "theorems"—that is, they are provable within alphabetics. Moreover, *only* statements of weight 6 are theorems. Suppose, now, we consider some alphabetic statement, S, which has numerical weight s. (So S is a string of A's, B's, and C's, while s is just a number.) Then the statement "S is provable in alphabetics"—which is a "meta-alphabetical" statement—is equivalent to the statement "$s = 6$"—which is a statement about numbers. That is, we have mapped a meta-alphabetical statement into an *arithmetical* statement in such a way that if one is true then so is the other.

Now, what Gödel did was something similar but vastly more difficult. He took the formal system consisting of logic and arithmetic—which is much more complicated than our silly made-up example of alphabetics—and invented a way of assigning numbers to the statements of this system, just as we assigned "numerical weights." This allowed him to map meta-arithmetical statements into arithmetical statements, in such a way that if one is true then so is the other. Then he was able, in an exceedingly clever way, to find a particular meta-arithmetical statement, "G is unprovable in arithmetic," which got mapped into the arithmetical statement G itself!

Putting It All Together

Now let us put this all together. Let **F** be a system containing arithmetic and simple logic. And let us only assume that this system is consistent. (We will not need the stronger statement that **F** is "sound.") Let arithmetical statements be denoted by capital letters, and the meta-arithmetical statements into which they are mapped be denoted by the same letters with "primes" after them. That way we will not get them confused. Let us consider the arithmetical statement G, and the meta-arithmetical statement into which it is mapped, G'. Suppose G is so cleverly constructed that G' is the statement "G is not provable in **F**." Of course, the arithmetical statement ¬G (which says the opposite of G) will be mapped into ¬G', which says "G *is* provable in **F**."

Suppose that G is provable in **F**. But that is just what ¬G' says. That would mean that the meta-arithmetical statement ¬G' is *true*. But if ¬G' is a true statement about statements, then ¬G (which it is mapped into) is a true statement about numbers, because they are yoked together by the mapping. So far we have that ¬G is a true statement about numbers; but is it provable in **F**?

Not all true statements about numbers are necessarily provable in **F**. However, there are some kinds of statements about numbers that if true are necessarily provable in **F** also. For example, consider the statement, "There are no prime numbers between 23 and 29." If that is true, then I can certainly prove it using the methods of arithmetic and logic. All I have to do is check all the numbers, 24, 25, 26, 27, and 28, and show that each is not prime. What Gödel was able to do was show that the statement ¬G *is* the kind of statement about numbers which if true is also provable in **F**. That brings us to the result that ¬G is provable in **F**. So now we have that ¬G is provable in **F**, and also that G is provable in **F**—which was assumed at the beginning of the previous paragraph. But that both G and ¬G are provable in **F** contradicts the assumption that **F** is consistent. Having run into a contradiction, we know that one of our assumptions must have been wrong. Therefore, either **F** is *not* consistent, or G is not provable in **F**.

It follows that if **F** *is* consistent, then G is not provable in **F**, which in turn means that G' is *true*. By the yoking, this means that G is also true. Therefore, if **F** is consistent there are propositions (among others G) which are not provable within **F** but which we can nevertheless see to be true.

Notes

Chapter 2. Materialism as an Anti-Religious Mythology

1. I am summarizing in my own words views that can be found expressed in countless books and articles.

2. Stanley L. Jaki, *Bible and Science* (Front Royal, Va.: Christendom Press, 1996), 57, 71.

3. Joseph Card. Ratzinger, *In the Beginning*, trans. Boniface Ramsey, O.P. (Grand Rapids, Mich.: William B. Eerdmans Publishing Company, 1995), 13–14. For an excellent discussion of the stark contrast between the religion of ancient Israel and the pagan religions of neighboring peoples and of the ancient Near East generally, see Nahum M. Sarna, *Understanding Genesis* (New York: Schocken Books, 1970).

4. St. Thomas Aquinas, II *Sententiarum* 12,3,1.

5. The views of the church fathers on the Hexahemeron are reviewed in appendix 7 of volume X of the Blackfriars edition of the *Summa Theologiae* of St. Thomas Aquinas (London: Eyre and Spottiswoode, Ltd., 1967), 203–4.

6. St. Thomas Aquinas, II *Sententiarum* 12,3,1.

7. St. Thomas Aquinas, *Summa Theologiae* Ia,68,1. St. Thomas is citing St. Augustine's *De Genesi ad Litteram* I,18,19,21 (Patrologiae Cursus Completus, Series Latina, ed. J. P. Migne, 34, 260–62).

8. St. Augustine, *On the Literal Meaning of Genesis* (*De Genesi ad Litteram*), trans. John Hammond Taylor, in *Ancient Christian Writers: The Works of the Fathers in Translation*, no. 41 (New York: Newman Press, 1982), 1: 42–43.

9. Jerome J. Langford, *Galileo, Science and the Church* (Ann Arbor: University of Michigan Press, 1971).

10. Quoted in Giorgio de Santillana, *The Crime of Galileo* (Chicago: University of Chicago Press, 1955), 99–100.

11. The most important of these can be found in the standard compendium of doctrinal statements of the Catholic Church, the *Enchiridion Symbolorum, Declarationem, et Definitionem*, ed. Heinrich Denziger and Adolf Schönmetzer (Barcelona: Herder, 1976). This edition is hereafter abbreviated as DS.

12. There is a long history of ideologically motivated attempts to find other theories or other scientists persecuted by the Catholic Church to place as exhibits alongside of Galileo. Some leading examples: (a) Roger Bacon (1214–1294). Bacon was a Franciscan monk, philosopher, and man of science. For some period of time between 1277 and 1292 he was imprisoned by his order, for reasons that are not entirely clear, but which,

according to the 1957 edition of *The Encyclopedia Britannica,* "could not have been of a scientific nature" and may have been largely caused by "his obnoxious attacks on his contemporaries." (b) Giordano Bruno (1548–1600). Bruno was burned at the stake on February 18, 1600. Although he advocated the Copernican system and held the bold view for that time that there are an infinity of inhabited worlds, it appears to have been his philosophical and theological teachings that led to his condemnation, rather than his scientific speculations. (c) Bernhard Bolzano (1781–1848). Bolzano, one of the pioneers of nineteenth-century mathematics, was deposed from the chair of the philosophy of religion at the University of Prague on December 24, 1819. However, it was the state not the church that was behind his deposition, and this was not for his mathematical, scientific, or even theological ideas but for his political and social views. The imperial government in Vienna objected to Bolzano's views on war, social rank, and civic obedience, and ordered therefore that he be charged with heresy. However, the ecclesiastical authorities consistently defended Bolzano as "an orthodox Catholic." Nevertheless, Bolzano eventually was deposed by imperial decree. See the biographical introduction by Donald Steele, S.J., in Bernard Bolzano, *Paradoxes of the Infinite* (London: Routledge and Kegan Paul, 1950). (d) Pierre Teilhard de Chardin (1881–1955). Teilhard was a paleontologist and Jesuit priest. On June 30, 1962, the Vatican issued a *monitum* which warned against uncritical acceptance of his theories (*Acta Apostolicae Sedis* 54 [1962] 526). Some have claimed that it was Teilhard's belief in biological evolution that brought him under suspicion by the church. The truth is that it was his theological and philosophical speculations. For a penetrating analysis of Teilhard's theology, which makes a strong case that he was indeed heterodox in his religious views, see the appendix of *Trojan Horse in the City of God,* Dietrich von Hildebrand (Manchester, N.Y.: Sophia Institute Press, 1993). It should be noted that at no time did the Catholic Church ever express disapproval of the theory of evolution. The only official statements have been by Pope Pius XII in 1950, in *Humani Generis,* stating that biological evolution, even the evolution of the human body from the bodies of lower organisms, as a scientific hypothesis was not in itself contrary to Catholic teaching, and by Pope John Paul II acknowledging that evolution is now more than merely a hypothesis.

  13. A. C. Crombie, *Robert Grosseteste and the Origins of Experimental Science* (Oxford: Clarendon Press, 1953). See also A. C. Crombie, *The History of Science from Augustine to Galileo* (New York: Dover, 1995), 2: 27–38.

  14. Crombie, *Robert Grosseteste and the Origins of Experimental Science,* 178–80; Crombie, *The History of Science from Augustine to Galileo,* 2: 69–72; Carl B. Boyer, *A History of Mathematics* (New York: John Wiley and Sons, Inc., 1968), 288–89.

  15. Crombie, *The History of Science from Augustine to Galileo,* 2: 87–97, 102–7; Marshall Clagett, *The Science of Mechanics in the Middle Ages* (Madison: University of Wisconsin Press, 1959); Boyer, *A History of Mathematics,* 289–95; *New Catholic Encyclopedia* (New York: McGraw-Hill, 1967).

  16. Thomas P. McTighe, "Nicolas of Cusa as a Forerunner of Modern Science," *Proceedings: 10th International Congress on the History of Science* (Ithaca: Cornell University Press, 1962), 619–22; Charles W. Misner, Kip S. Thorne, and John Archibald Wheeler, *Gravitation* (San Francisco: W. H. Freeman, 1973), 754; Frederick Copleston, S.J., *A History of Philosophy* (New York: Doubleday, 1963), 3, II, ch. 15.

  17. *New Catholic Encyclopedia.*

18. R. N. Wilson, *Reflecting Telescope Optics*, vol. I, *Basic Design Theory and Its Historical Development* (Berlin/Heidelberg: Springer-Verlag, 1996), 3–4, 469.

19. Johann Schreiber and William F. Rigge, "Jesuit Astronomy," in *Popular Astronomy* 12 (1904).

20. *Dictionary of Scientific Biography* (New York: Scribners, 1970); *Catholic Encyclopedia* (New York: Encyclopedia Press, Inc., 1913).

21. Ibid.

22. Ibid.

23. Peter Doig, *A Concise History of Astronomy* (New York: Philosophical Library, 1951), 298; *New Catholic Encyclopedia*.

24. Anton Pannekoek, *A History of Astronomy* (New York: Interscience Publishers, 1961), 352–53; Harlow Shapley and Helen E. Howarth, *Source Book in Astronomy* (Cambridge: Harvard University Press, 1960), 180–82.

25. *Dictionary of Scientific Biography*.

26. Robin Marantz Henig, *The Monk in the Garden: The Lost and Found Genius of Gregor Mendel* (New York: Houghton-Mifflin, 2000).

27. A. H. Broderick, *Father of Prehistory, the Abbé Henri Breuil: His Life and Times* (New York: Morrow, 1963).

28. Andrei Deprit, "Monsignor Georges Lemaître," in *The Big Bang and Georges Lemaître*, ed. André Berger (Dordrecht: D. Reidel Publishing Company, 1984), 363–92; Misner, Thorne, and Wheeler, *Gravitation*, 758.

29. Ralph F. Wolf, "Rich Man, Poor Man," *Scientific Monthly* (Feb. 1952): 69–75; *New Catholic Encyclopedia*; J. A. Nieuwland, "Synthetic Rubber from a Gas," *Scientific American* (Nov. 1935) 262ff.

30. Boyer, *A History of Mathematics*; Bolzano, *Paradoxes of the Infinite*.

31. Stephen Jay Gould, *Full House: The Spread of Excellence from Plato to Darwin* (New York: Harmony Books, 1996), 28.

32. *Enchiridion Symbolorum, Declarationem, et Definitionem*, DS 3009, 3010.

33. Ibid., DS 3033.

34. The First Vatican Council formally condemned the proposition that "the One true God, our Creator and Lord, cannot be known with certainty with the natural light of human reason through the things that are created." *Enchiridion Symbolorum, Declarationem, et Definitionem*, DS 3026.

35. John Calvin, *Institutes of the Christian Religion*, trans. John Allen (Philadelphia: Presbyterian Board of Publications, 1928), bk. I, ch. VIII.

36. Avery Dulles, S.J., *The Assurance of Things Hoped For: A Theology of Christian Faith* (Oxford: Oxford University Press, 1994). Chapter 10 deals with the grounding of faith. John Henry Card. Newman, *An Essay in Aid of a Grammar of Assent* (Notre Dame, Ind.: University of Notre Dame Press, 1979); *Catechism of the Catholic Church* (Vatican City: Libreria Editrice Vaticana, 1994), secs. 150–60; Pope John Paul II, encyclical letter *Fides et ratio* ("Faith and Reason").

37. St. Irenaeus of Lyons, *Adversus Haeresis* 2,9,1, English translation in William A. Jurgens, *The Faith of the Early Fathers* (Collegeville, Minn.: Liturgical Press, 1970), 86.

38. Calvin, *Institutes of the Christian Religion*, 58.

39. Bernard J. F. Lonergan, S.J., *Insight: A Study of Human Understanding* (New York: The Philosophical Library, 1967), 639–77.

40. John Maddox, *What Remains to Be Discovered: Mapping the Secrets of the Universe, the Origins of Life, and the Future of the Human Race* (New York: The Free Press, Simon and Schuster, Inc., 1998), 281.

41. David J. Chalmers, *The Conscious Mind: In Search of a Fundamental Theory* (Oxford: Oxford University Press, 1996), 162.

42. Avshalom C. Elitzur, "Consciousness and the Incompleteness of the Physical Explanation of Behavior," *The Journal of Mind and Behavior* 10 (1989): 1–20.

43. W. V. O. Quine, *Word and Object* (Cambridge, Mass.: M.I.T. Press, 1960), 264; W. V. O. Quine "Minds and Verbal Dispositions," in *Mind and Language, Wolfson College Lectures, 1974* (Oxford: Clarendon Press, 1975), 83–95.

44. The idea that the "self" does not exist goes back at least to David Hume, *A Treatise of Human Nature*, ed. David Fate Norton and Mary J. Norton (Oxford: Oxford University Press, 2000), 164–71. It can be found expressed or implied in many recent books about the mind written from a materialist point of view, where the self is regarded as a kind of fictitious actor in the internal "scenarios" that make up what we call consciousness.

Chapter 3.   Scientific Materialism and Nature

1. In the words of E. O. Wilson, "Virtually all contemporary scientists and philosophers expert on the subject agree that the mind, which comprises consciousness and rational process, is the brain at work." E. O. Wilson, *Consilience: The Unity of Knowledge* (New York: Alfred A. Knopf, 1998), 98. The prevailing view among philosophers of mind and cognitive scientists is that consciousness, to the extent that it has any reality at all, is merely a reflection of physical processes in the brain, and has no causal influence of its own at all. Consciousness is something entirely passive. This view is so widespread that Avshalom Elitzur, a philosopher and scientist who dissents from it, says, "I think one may talk here about the dogma of passivity." Elitzur, "Consciousness and the Incompleteness of the Physical Explanation of Behavior." For a review of contemporary thinking about the mind, see Chalmers, *The Conscious Mind: In Search of a Fundamental Theory.*

2. In 1997, the chess program Deep Blue defeated world champion Garry Kasparov 3.5–2.5 in a six-game match. For an expert discussion of the match and the current chess abilities of computers, see Boris Gulko, "Is Chess Finished?" in *Commentary* (July 1997), 45.

3. According to Democritus of Abdera and the Atomist school of philosophy, nothing is real except "atoms and the void." Copleston, *A History of Philosophy*, 1, I, 144–47.

4. See, for example, Gould, *Full House*, 17–20.

5. There are about 10^{80} protons and neutrons in the observable universe, and about 4×10^{51} in the earth, giving a ratio of about 2.5×10^{28} to 1. The earth's oceans have about 1.4×10^{24} cm^3 of water, which is in the same ratio to a droplet of 0.06 mm^3.

6. Bertrand Russell, *Religion and Science* (Oxford: Oxford University Press, 1961), 222.

7. Dirac's comments in this paragraph are quoted in Abraham Pais, *Inward Bound: Of Matter and Forces in the Physical World* (Oxford: Oxford University Press, 1986), 290–91.

8. Hermann Weyl, *The Open World: Three Lectures on the Metaphysical Implications of Science* (New Haven, Conn.: Yale University Press, 1932), 54–55.

Chapter 4.   The Expectations

1. St. Thomas Aquinas, *Summa Contra Gentiles*, bk. II, ch. 38, trans. James F. Anderson (South Bend, Ind.: University of Notre Dame Press, 1975), 2: 112–14.

2. The Fourth Lateran Council declared that "from the beginning of time [God] formed created things out of nothing" (*"ab initio temporis utramque de nihilo condidit creaturam"*), *Enchiridion Symbolorum, Declarationem, et Definitionem*, DS 800, DS 3002.

3. The word *eternal* is commonly used to mean "of infinite duration." However, it can also mean "timeless" or "outside of time," and does so when applied to God. For a good discussion of the two senses of the word see Mortimer Adler, *The Great Ideas, A Lexicon of Western Thought* (New York: Macmillan Publishing Company, 1952), 194–201.

4. Although it is conventional to assert that modern physics has dispensed with final causes, this is debatable. There are many cases in modern physics where the simplest way to analyze a physical situation or to get the answer to a physics problem is by reference to a future state or states of the system. This is true, for instance, of systems that are approaching an equilibrium state. (For example, a hot cup of coffee will cool until it reaches the same temperature as the circumambient air.) More generally, the laws of classical physics can be formulated as "action principles," according to which any physical system will take the path leading from its initial state to its final state, which has the least "action." The analogous formulation of quantum theory is the so-called "path integral."

5. P. C. W. Davies, *God and the New Physics* (New York: Simon and Schuster, Inc., 1983), 42.

Chapter 6.   The Big Bang

1. This is a common misunderstanding of the expanding universe. The well-known science fiction writer Poul Anderson based an entire book, *Tau Zero*, on this misconception.

2. There is a subtlety here. In a "closed universe," where the volume of the universe is always finite, the total volume of the universe gets smaller as one goes back toward the Big Bang, and is zero at the first instant. In an "open universe" the volume of the universe is infinite at all times. It is less confusing, in this case, to talk about a finite patch of the universe rather than the whole thing. The volume of any finite patch of an open universe gets smaller as one goes back toward the Big Bang, and is zero at the Big Bang.

3. Abraham Pais took the title of his remarkable scientific biography of Einstein from this saying. Abraham Pais, *Subtle Is the Lord: The Science and Life of Albert Einstein* (Oxford: Oxford University Press, 1982).

4. Misner, Thorne, and Wheeler, *Gravitation*, 759.

5. In 1917, Willem de Sitter found a solution of Einstein's equations that describes an exponentially expanding universe. This solution is not relevant to describing the cosmic expansion discovered by Slipher, Hubble, and Humason. However, it does describe the expansion that occurs during the "inflationary" phases of the early universe, according to most present models of inflation. (See pp. 54–57) Moreover, if it turns out

that there is a small cosmological constant, as some recent evidence suggests, then eventually the expansion of the universe will approximate more and more closely the de Sitter model.

6. Robert Jastrow, *God and the Astronomers* (New York: Norton, 1992), 32.

7. These expressions of distaste for the idea of a cosmic beginning are reported by Jastrow in *God and the Astronomers,* the Einstein quote on p. 21 and the Eddington and Nernst quotes on p. 104.

8. A. Linde, D. Linde, and A. Mezhlumian, "From the Big Bang to the Theory of a Stationary Universe," *Physical Review* D49 (1994): 1783.

9. Stephen G. Brush, "How Cosmology Became a Science," *Scientific American* (Aug. 1992), 62–70.

10. It should be noted that Eddington was a practicing Quaker. Presumably his initial distaste for the Big Bang Theory was based on other grounds than hostility to religion.

11. The radiation in the early, "radiation dominated" period of the universe's history was not all in the form of light. Much of it was in the form of neutrinos, and, earlier, in the form of other particles as well.

12. There are some who continue to argue against the Big Bang Theory. See Eric J. Lerner, *The Big Bang Never Happened* (New York: Times Books/Random House, 1991). However, the scientific evidence for the Big Bang is overwhelming.

Chapter 7.   Was the Big Bang Really the Beginning?

1. St. Augustine is commonly misquoted as saying, in answer to the question of what God was doing before he created Heaven and Earth, that he was creating Hell for those who ask such questions. This mistake is repeated, for instance, by Robert Jastrow in *God and the Astronomers.* This is what St. Augustine actually said: "I do not give the answer that someone is said to have given, evading by a joke the force of the objection: 'He was preparing hell,' he said, 'for those prying into such deep subjects.' . . . I do not answer in this way. I would rather respond, 'I do not know,' concerning what I do not know than say something for which a man inquiring about such profound matters is laughed at, while the one giving a false answer is praised." *Confessions,* trans. John K. Ryan (New York: Doubleday and Co., Inc., 1960), bk. 11, ch. 12.

2. St. Augustine, *Confessions,* bk. 11, ch. 30.

3. Ibid., bk. 11, ch. 13.

4. Ibid.

5. As noted by Steven Weinberg, "Book XI of Augustine's *Confessions* contains a famous discussion of the nature of time, and it seems to have become a tradition to quote from this chapter in writing about quantum cosmology." *Reviews of Modern Physics* 61 (1989): 15, n15.

6. Bertrand Russell, *History of Western Philosophy* (London: Allen and Unwin, 1946), 373.

7. Most of the philosophers of antiquity taught that the world was eternal. This was certainly true of Aristotle, and probably true of Plato. While Plato in the *Timaeus* speaks of the Demiurge forming the world, many scholars doubt that he meant this myth to refer

to an actual creation in time, and in any event he believed that the matter from which the world is formed is eternal. Some of the later Platonists, such as Plutarch, did believe that the world had a beginning in time. But these thinkers were essentially monotheists. Philo of Alexandria, a Jewish philosopher of the early first century, spoke of time having a beginning, as did some other neo-Platonists, but it is not until St. Augustine that the implications of this were fully appreciated and we find such statements as "There was no 'then'" before the world began.

8. J. Khoury, B. A. Ovrut, N. Seiberg, P. J. Steinhardt, and N. Turok, "From Big Crunch to Big Bang," *Physical Review* D65 (2002) 086702; P. J. Steinhardt and N. Turok, "Cosmic Evolution in a Cyclic Universe," *Physical Review* D65 (2002) 126003.

9. A universe that is inflating with constant Hubble parameter has a space-time described by the de Sitter solution of Einstein's equations. But the complete de Sitter solution can be shown actually to describe a universe that contracts from $t = $ -infinity until some finite time, and then expands again to $t = $ +infinity. Such a solution, while it has no beginning in time, would be regarded as "unphysical" by most physicists. It is therefore doubtful that "eternal inflation" can have been going on without a beginning.

Chapter 8. What If the Big Bang Was Not the Beginning?

1. Even the fabric of space shows the effects of the passage of time. The geometry of space-time itself has entropy associated with it. An interesting technical discussion of this is Roger Penrose, "Singularities and Time-asymmetry," in *General Relativity: An Einstein Centenary Survey* (Cambridge: Cambridge University Press, 1979), 612–13, 629–30.

2. Ratzinger, *In the Beginning*, 22.

Chapter 9. The Argument from Design

1. See footnote 3 of chapter 2.

2. Jacob Neusner, *Confronting Creation: How Judaism Reads Genesis, an Anthology of Genesis Rabbah* (University of South Carolina Press, 1991), 15.

3. Genesis Rabbah Parashiyyot I.I.2, in Neusner, *Confronting Creation*, 15, 26.

4. Neusner, *Confronting Creation*, 15.

5. Wilson, *Consilience: The Unity of Knowledge*, 31.

6. See, for example, the Epistle to the Corinthians, written ca. 96 A.D., by Clement of Rome, especially chapters 19 and 20, in *Early Christian Writings: The Apostolic Fathers*, ed. Betty Radice, trans. Maxwell Staniforth (New York: Penguin, 1968).

7. Minucius Felix, *Octavius* 18,4, in Jurgens, *The Faith of the Early Fathers*, 1: 109.

8. Calvin, *Institutes of the Christian Religion*, 58.

9. William Paley, *The Works of William Paley* (Oxford: Clarendon Press, 1938), 4: 1.

10. Lord Macaulay, *Critical and Historical Essays* (London: Longman, Green and Company, 1898), 2: 280.

11. *Webster's Seventh New Collegiate Dictionary* (Springfield, Mass.: G. and C. Merriam Company, 1970).

12. Calvin, *Institutes of the Christian Religion*, 58.

13. Ibid., 59.

Chapter 10.   The Attack on the Argument from Design

1. Daniel C. Dennett, *Darwin's Dangerous Idea: Evolution and the Meanings of Life* (New York: Simon and Schuster, 1995).
2. Richard Dawkins, *The Blind Watchmaker: Why the Evidence of Evolution Reveals a Universe Without Design* (New York: W. W. Norton, 1996), 6.
3. Dawkins, *The Blind Watchmaker*.
4. Richard Dawkins, *Climbing Mount Improbable* (New York: W. W. Norton, 1996), 6.
5. Gretchen Vogel, "Finding Life's Limits," in *Science* 282 (1998), 1399. This article reports on the conclusions reached by scientists at the Workshop on Size Limits of Very Small Microorganisms, held 22–23 October 1998.
6. William A. Dembski in "Science and Design," *First Things* 86 (October 1998): 21.

Chapter 11.   The Design Argument and the Laws of Nature

1. James R. Newman, "Laplace," in *Lives in Science, a Scientific American Book* (New York: Simon and Schuster, 1957), 51. According to the *New Catholic Encyclopedia*, Laplace died a Catholic.
2. Banesh Hoffmann, *Albert Einstein: Creator and Rebel* (New York: Viking Press, 1972), 141.

Chapter 13.   "What Immortal Hand or Eye?"

1. Anthony Zee, *Fearful Symmetry: The Search for Beauty in Modern Physics* (New York: Macmillan, 1986), 280–1.
2. Theorists have discovered that superstring theories are really limiting cases of something they call "M-theory." I will nevertheless keep using the term *superstring theory* for it in the text.
3. John Horgan, *The End of Science* (New York: Addison-Wesley, 1996), 69.
4. David Hume, *Dialogues Concerning Natural Religion*, ed. N. Pike (Indianapolis, Ind.: Bobbs-Merrill, 1970), 50.
5. Weyl, *The Open World*, 28–29.
6. Stephen Jay Gould, *Wonderful Life: The Burgess Shale and the Nature of History* (New York: W. W. Norton, 1989).
7. Michael J. Behe, *Darwin's Black Box: the Biochemical Challenge to Evolution* (New York: The Free Press, 1997).
8. Werner Heisenberg, *Across the Frontiers*, trans. Peter Heath (New York: Harper and Row, 1974), 37.
9. Dawkins, *The Blind Watchmaker*.

Chapter 14.   The Expectations

1. Epistle to Diognetus, 10, in *Early Christian Writings: The Apostolic Fathers*, ed. Betty Radice, 181.
2. Victor Stenger, *Not by Design: The Origin of the Universe* (Buffalo, N.Y.: Prometheus Books, 1988), 12.

3. Steven Weinberg, *The First Three Minutes: A Modern View of the Origin of the Universe* (Glasgow: William Collins, 1977), 148.

4. Richard Dawkins, "Science and God: A Warming Trend?," *Science* 277 (1997), 890.

5. Gould, *Full House*, 17.

6. Ibid., 18.

7. Ibid., 14, 18, 29, 41.

8. Brandon Carter, in *Confrontation of Cosmological Theories with Observation*, ed. M. S. Longair (Dordrecht, Netherlands: Reidel, 1974), 291.

Chapter 15.   The Anthropic Coincidences

1. Those interested in reading more about anthropic coincidences can turn to J. D. Barrow and F. J. Tipler, *The Cosmological Anthropic Principle* (Oxford: Oxford University Press, 1986), and P. C. W. Davies, *The Accidental Universe* (Cambridge: Cambridge University Press, 1982).

2. Barrow and Tipler, *The Cosmological Anthropic Principle*, ch. 5, sec. 5.5.

3. The twenty-five elements that are used in the human body are, with their atomic numbers, hydrogen (1), boron (5), carbon (6), nitrogen (7), oxygen (8), fluorine (9), sodium (11), magnesium (12), silicon (14), phosphorus (15), sulfur (16), chlorine (17), potassium (19), calcium (20), vanadium (23), chromium (24), manganese (25), iron (26), cobalt (27), copper (29), zinc (30), selenium (34), molybdenum (42), tin (50), and iodine (53).

4. The material in this section is based on Barrow and Tipler, *The Cosmological Anthropic Principle*, 322.

5. V. Agrawal, S. M. Barr, J. F. Donoghue, and D. Seckel, "Viable Range of the Mass Scale of the Standard Model," *Physical Review* D57 (1998): 5480; "Anthropic Considerations in Multiple-domain Theories of the Scale of Electroweak Symmetry Breaking," *Physical Review Letters* 80 (1998): 1822.

6. T. E. Jeltema and M. Sher, "The Triple-alpha Process and Anthropically Allowed Values for the Weak Scale," *Physical Review* D61 (2000): 017301.

7. Steven Weinberg has suggested that the value of the cosmological constant may be explained by the Weak Anthropic Principle. S. Weinberg, "Anthropic Bound on the Cosmological Constant," *Physical Review Letters* 57 (1987): 2607.

8. I am defining the spatial curvature to be the ratio of the square of the horizon length to the square of the radius of curvature of space.

9. The material in this section is based on Barrow and Tipler, *The Cosmological Anthropic Principle*, ch. 4, sec. 4.8.

Chapter 16.   Objections to the Idea of Anthropic Coincidences

1. Communication from an editor of *Physical Review* to authors of first paper cited in note 5 of chapter 15.

2. Brandon Carter, in *Confrontation of Cosmological Theories with Observation*, 291.

3. A. D. Sakharov, Nobel lecture, read by Yelena Bonner at Nobel Peace Prize ceremony, Oslo, Norway, Dec. 11, 1975.

4. B. J. Carr and M. J. Rees, "The anthropic principle and the structure of the physical world," *Nature* 278 (1979), 605.

5. Yakov B. Zel'dovich, *Pis'ma v A. Zh.* (*Letters to the Astrophysical Journal*) 7 (1981): 579.

6. Barrow and Tipler, *The Cosmological Anthropic Principle.*

7. S. Weinberg, *Physical Review Letters* 59 (1987): 2607.

8. L. Okun, "Fundamental Constants in Physics," in *Sakharov Memorial Lectures in Physics*, ed. L. V. Kaldyshand and V. Ya. Fainberg (New York: Nova Science Publishers, Inc., 1992), 2: 819–40.

9. A. Linde, *Particle Physics and Inflationary Cosmology* (New York: Gordon and Breach, 1990).

10. A. Vilenkin, "Open Universes, Inflation, and the Anthropic Principle," *International Journal of Theoretical Physics* 38 (1999): 3135; J. Garriga, T. Tanaka, and A. Vilenkin, "Density Parameter and the Anthropic Principle," *Physical Review* D60 (1999): 023501.

11. N. Turok and S. Hawking, *Physics Letters* B432 (1998): 271.

12. G. Feinberg and R. Shapiro, *Life Beyond Earth* (New York: William Morrow, 1980).

13. The minimal supersymmetric SU(5) model.

14. B. J. Carr and M. J. Rees, *Nature* 278 (1979), 605.

Chapter 17. Alternative Explanations of the Anthropic Coincidences

1. Barrow and Tipler, *The Cosmological Anthropic Principle.*

2. V. Agrawal, S. M. Barr, J. F. Donoghue, and D. Seckel, *Physical Review Letters* 80 (1998): 1822; *Physical Review* D57 (1998): 5480.

3. David Kellogg Lewis, *On the Plurality of Worlds* (Oxford: Oxford University Press, 1986).

4. The philosopher John Leslie has argued along similar lines. J. Leslie, *Universes* (London: Routledge, 1989).

Chapter 18. Why Is the Universe So Big?

1. Blaise Pascal, *Pensées*, trans. T. S. Eliot (New York: E. P. Dutton and Co., Inc., 1958), 61.

2. Bertrand Russell, "My Mental Development," in *The Philosophy of Bertrand Russell*, ed. Paul Arthur Schilpp (Evanston, Ill.: Northwestern University Press, 1944), 19–20.

3. Quoted in J. C. Eccles and K. R. Popper, *The Self and Its Brain* (New York: Springer, 1977), 61.

4. Copleston, *A History of Philosophy*, 3, II: 47–48.

5. The arguments in this section are taken from B. J. Carr and M. J. Rees, *Nature* 278 (1979), 605.

6. α_G is defined to be Newton's constant times the square of the proton mass.

Chapter 19. The Issue

1. St. Irenaeus of Lyons, *Adversus Haeresis* 4,4,3.

2. Calvin, *Institutes of the Christian Religion*, 176.

3. *Catechism of the Catholic Church*, sec. 1705.

4. Ibid., sec. 33.

5. Simon Easteal, Chris Collet, and David Betty, *The Mammalian Molecular Clock* (New York: Springer-Verlag, 1990).

6. Eccles and Popper, *The Self and Its Brain*, 338.

7. Malcolm W. Browne, "Who Needs Jokes?: Human brain has a ticklish spot," in *The New York Times*, March 10, 1998, sec. F, p. 1.

8. Pierre Simon Marquis de Laplace, *A Philosophical Essay on Probabilities* (New York: Dover, 1951), 4–5.

9. Since I say that physical determinism is incompatible with human freedom, it may be wondered why divine foreknowledge is not also so. The important point is that physical determinism implies that a human decision can be completely understood solely in terms of what the material constituents of the human body were doing before the decision was made and without reference to the human "will." But the fact that God "foreknows" a human decision does not make it unfree, since what God knows is precisely that the decision will be made freely. That is, God foreknows our acts as what they truly are, acts of a free will. As St. Augustine pointed out, God's knowledge of future human decisions no more makes them unfree than our own knowledge of past human decisions makes them unfree. See St. Augustine, *On the Free Choice of the Will*, trans. Anna S. Benjamin and L. H. Hackstaff (New York: The Bobbs-Merrill Co. Inc., 1964).

10. Roger Penrose, *Shadows of the Mind: The Search for the Missing Science of Consciousness* (Oxford: Oxford University Press, 1994), 202.

11. *Catechism of the Catholic Church*, sec. 365.

Chapter 20.   Determinism and Free Will

1. The many-worlds interpretation (MWI) of quantum theory will be explained in chapter 25. In a certain sense the MWI makes quantum theory deterministic. There is no unpredictable "collapse of the wavefunction" (see chapter 24, esp. pp. 236–37). Rather, the wavefunction at one time uniquely determines it at all later times. Some have suggested that quantum theory even in the MWI allows free will. The point is that the unique wavefunction has a multiplicity of "branches" corresponding to different paths the world can take. When faced with a choice of alternatives, "I" split up into many alter egos, each of whom goes down a different branch and experiences a different one of the alternatives. Perhaps, it is suggested, I can decide which branch "I" will experience—i.e., which alter ego "I" will be. I do not regard this as genuine free will. But if it is, that would only strengthen the case I am making here that quantum theory creates an opening for free will, since it would no longer rely upon the traditional interpretation of quantum theory.

2. John R. Lucas, *The Freedom of the Will* (Oxford: Oxford University Press, 1970); Eccles and Popper, *The Self and Its Brain*; David Hodgson, *The Mind Matters: Consciousness and Choice in a Quantum World* (Oxford: Oxford University Press, 1988); Michael Lockwood, *Mind, Brain, and Quantum: The Compound 'I'*, (Oxford: Blackwell, 1989); Euan Squires, *Conscious Mind in the Physical World* (London: Adam Hilger, IOP

Publishing, Ltd., 1990); H. P. Stapp, *Mind, Matter, and Quantum Mechanics* (New York: Springer-Verlag, 1993).

3. Chalmers, *The Conscious Mind*, 157.

4. Penrose, *Shadows of the Mind*, 350.

5. Ibid., 348–49, 367–69.

6. F. Beck and J. C. Eccles, "Quantum aspects of consciousness and the role of consciousness," *Proceedings of the National Academy of Sciences* 89 (1992): 11357–61; J. C. Eccles, *How the Self Controls Its Brain* (Berlin: Springer-Verlag, 1994).

7. Weyl, *The Open World*, 54–55.

8. John Horgan, *The Undiscovered Mind: How the Human Brain Defies Replication, Medication, and Explanation* (New York: Touchstone, 1999), 247.

9. Wilson, *Consilience*, 119.

10. Daniel C. Dennett, *Consciousness Explained* (Boston: Little, Brown, 1991).

11. Gould, *Full House*, 28.

12. Quoted in "Science and Religion: Bridging the great divide," George Johnson, *The New York Times*, June 30, 1998, sec. F, p.4.

13. James Boswell, *Life of Johnson*, ed. G. B. Hill, revised by L. F. Powell (Oxford: Oxford University Press, 1934), 2: 82.

14. Ibid., 4: 239.

Chapter 21.   Can Matter "Understand"?

1. Mortimer J. Adler, *Intellect: Mind over Matter* (New York: Macmillan Publishing Co., 1990), ch. 4.

2. According to Aristotle the human mind, in contrast to the minds of animals, had an immaterial component, which he called the active intellect (Nous). "The active intellect abstracts forms from the [mental] images or phantasmata, which, when received in the passive intellect, are actual concepts." Copleston, *A History of Philosophy*, 1, II, 71. The immateriality of the active intellect has been held by many philosophers ancient, mediaeval, and modern.

3. John McCarthy, "Ascribing Mental Qualities to Machines," in *Philosophical Perspectives on Artificial Intelligence*, ed. Martin Ringle (Atlantic Highlands, N.J.: Humanities Press, 1979).

4. Francis Crick, *The Astonishing Hypothesis: The Scientific Search for the Soul* (New York: Charles Scribner's Sons, 1994), 3.

5. Maddox, *What Remains to Be Discovered*, 281.

6. George Johnson, "Does the Universe Follow Mathematical Law?" *The New York Times*, Feb. 10, 1998.

7. Stanislas Dehaene, *The Number Sense: How the Mind Creates Mathematics* (Oxford: Oxford University Press, 1997).

8. Eccles and Popper, *The Self and Its Brain*; K. R. Popper, *Knowledge and the Mind-Body Problem: In Defense of Interaction* (London: Routledge, 1994).

9. G. K. Chesterton, *Orthodoxy* (New York: Doubleday, 1959), ch. 3.

10. Sharon Begley, "Thinking Will Make It So," *Newsweek*, April 5, 1999, p. 64.

11. Weyl, *The Open World*, 31–32.

12. J. B. S. Haldane, *The Inequality of Man* (London: Chatto and Windus, 1932).

13. Chesterton, *Orthodoxy*, 33.

14. Stephen Hawking, *A Brief History of Time: From the Big Bang to Black Holes* (London: Bantam, 1988), 12.

15. Cyril Bailey, *Epicurus: The Extant Remains* (Oxford: Clarendon Press, 1926).

16. J. B. S. Haldane, "I repent an error," in *The Literary Guide* (April 1, 1954), 7, 29.

17. Penrose, *Shadows of the Mind*, 147–50.

18. Kitty Ferguson, *The Fire in the Equations: Science, Religion, and the Search for God* (Grand Rapids, Mich.: William B. Eerdmans Publishing Co., 1994), 63.

19. Arguments similar to the ones I am developing here have been put forward by Prof. Katherin Rogers of the University of Delaware.

20. St. Thomas Aquinas, *Summa Theologiae*, I, 24, 5.

21. G. H. Hardy, *A Mathematician's Apology* (Cambridge: Cambridge University Press, 1940), 70; quoted in Ferguson, *The Fire in the Equations*, 63.

22. Quoted in Weyl, *The Open World*, 10–11.

23. Penrose, *Shadows of the Mind*, 55–56.

24. St. Augustine, *On Free Choice of the Will*, bk. 2, ch. 8. The example St. Augustine used was not well chosen, being in essence that $a + a = 2 \times a$ for any a, which can be taken to be a definition of multiplication by 2. But if he had used $a + a = a \times 2$, or equivalently $2 \times a = a \times 2$, he would have been using the same example as Penrose, with b taken to be the specific number 2.

25. Sir Isaac Newton, *The Mathematical Principles of Natural Philosophy*, trans. Andrew Motte (New York: Daniel Adee, 1848), 507.

Chapter 22. Is the Human Mind Just a Computer?

1. I am being a little careless in stating these rules. In the axiom, x does not stand for any string of symbols without restriction. For example, x should not be something like $a = b$. Similarly, the rule of inference must be stated more carefully. For example, it should not allow x or y to stand for a fragment like ")".

2. Actually Gödel proved two closely related theorems, the first and second incompleteness theorems, which are sometimes referred to together as Gödel's Theorem. I shall not bother to distinguish between them. For a semi-popular account see Ernest Nagel and James R. Newman, *Gödel's Proof* (New York: New York University Press, 1958). I present a restatement of Nagel and Newman's explanation in appendix C.

3. In his original proof, Gödel had to use the somewhat stronger assumption of what is called "ω-consistency rather than mere consistency. However, in 1936 J. Barkley Rosser proved the same results using consistency.

4. This is Gödel's second theorem. See footnote 2 for this chapter.

5. John R. Lucas, "Minds, machines, and Gödel," *Philosophy* 36 (1961): 120, reprinted in *The Modeling of Mind*, ed. K. M. Sayre and F. J. Crossen (Notre Dame, Ind.: University of Notre Dame Press, 1963); Lucas, *The Freedom of the Will*.

6. Lucas, "Minds, machines, and Gödel," in *The Modeling of Mind*, 255.

7. Quoted in Penrose, *Shadows of the Mind*, 128.

8. Roger Penrose, *The Emperor's New Mind* (Oxford: Oxford University Press, 1989).

9. "Multiple Book Review of *The Emperor's New Mind: Concerning Computers, Minds, and the Laws of Physics*," *Behavioral and Brain Sciences* 13 (4), 1990, 643–705.

10. Penrose, *Shadows of the Mind*.

11. See especially "Penrose's Gödelian Argument: A Review of *Shadows of the Mind*," by Solomon Feferman, *Psyche* 2 (7), May 1995. Penrose replied to nine prominent critics in "Beyond the Doubting of a Shadow: A Reply to Commentaries on *Shadows of the Mind*," *Psyche* 2 (23), Jan. 1996.

12. Penrose argues that any explanation of the human mind must involve as-yet-unknown laws of physics and as-yet-unknown "non-computational" mechanisms. See *Shadows of the Mind*, part II, and especially sec. 4.2.

13. Penrose, *Shadows of the Mind*, 84–86.

14. Lucas, "Mind, machines, and Gödel," in *The Modeling of Mind*, 264.

15. Ibid., 266.

16. Chesterton, *Orthodoxy*, 32.

Chapter 23.   What Does the Human Mind Have That Computers Lack?

1. Penrose, *Shadows of the Mind*, 56.

2. Ibid., 109. Lonergan, *Insight: A Study of Human Understanding*, 13–14. It is interesting to compare what Penrose and Lonergan have to say on the subject of understanding what the natural numbers are.

3. Quoted in Stapp, *Mind, Matter, and Quantum Mechanics*, 11.

4. Lonergan, *Insight: A Study of Human Understanding*, 639–77.

5. The simplicity of God was defined as an article of Christian faith by the Fourth Lateran Council in 1215. The council declared that God is a *"substantia seu natura simplex omnino"* ("substance or nature entirely simple"). (*Enchiridion Symbolorum, Declarationem, et Definitionem*, DS 800, DS 3002.)

6. St. Irenaeus of Lyons, *Adversus Haeresis* 2,13,3, trans. Jurgens, *The Faith of the Early Fathers*, 87.

7. *Catechism of the Catholic Church*, sec. 365.

Chapter 24.   Quantum Theory and the Mind

1. Quantum electrodynamics is the most precisely tested theory in all of science. The "anomalous magnetic moment" of the electron has been calculated using quantum electrodynamics to be $a_e \times 10^{12} = 1159652201.2 +- 2.1 +- 27.1$, and measured in the laboratory to be $a_e \times 10^{12} = 1159652188.4 +- 4.3$. (The numbers after +- are the theoretical and experimental uncertainties.) This is an agreement to one part in a hundred million. That is equivalent to calculating the distance between a point in California and a point in New York and getting it right to within two inches. The anomalous magnetic moment is a purely quantum effect.

2. Quoted in Squires, *Conscious Mind in the Physical World*, 183.

3. P. C. W. Davies and J. R. Brown, *The Ghost in the Atom* (Cambridge: Cambridge University Press, 1986), 75.

4. Eugene P. Wigner, "Remarks on the Mind-Body Question," in *The Scientist Speculates*, ed. I. J. Good (London: William Heinemann, Ltd., 1961), reprinted in Eugene P. Wigner, *Symmetries and Reflections: Scientific Essays* (Woodbridge, Conn.: Ox Bow Press, 1979), 176.

5. John von Neumann, *Mathematical Foundations of Quantum Mechanics*, English translation (Princeton, N.J.: Princeton University Press, 1955).

6. Fritz London and Edmond Bauer, *La Théorie de l'Observation en Mècanique Quantique* (Paris: Herrmann and Cie., 1939).

7. Tom Banks, "The State of Matrix Theory," *Nuclear Physics* B (Proc. Suppl.) 62A-C (1998): 341.

8. Rudolf Peierls, "Observation in Quantum Mechanics and the 'Collapse of the Wavefunction,'" in *Symposium on the Foundations of Modern Physics*, ed. Pekka Lahti and Peter Mittelstaedt (Singapore: World Scientific Publishing Co., 1985), 193.

9. Davies and Brown, *The Ghost in the Atom*, 74.

10. In quantum theory one also has coordinates and equations of motion for them. However, they have a different status. The coordinates are operators in a Hilbert space, and the equations of motion are operator equations. Their connection to what is actually observed is not quite as straightforward as in classical physics.

11. In special cases, quantum probability amplitudes may correspond to a probability of 100 percent, which would be certainty, of course. I should therefore say that generally speaking the probability amplitudes given by the Schrödinger evolution do not correspond to definite outcomes.

12. Davies and Brown, *The Ghost in the Atom*, 74.

13. Peierls, "Observation in Quantum Mechanics and the 'Collapse of the Wavefunction,'" 193.

14. Wigner, *Symmetries and Reflections*, 172.

15. Davies and Brown, *The Ghost in the Atom*, 74.

16. Weyl, *The Open World*, 54–55.

17. Wigner, *Symmetries and Reflections*, 172.

18. Peierls, "Observation in Quantum Mechanics and the 'Collapse of the Wavefunction,'" 193.

19. I am indebted to a discussion with Peierls for giving me a better understanding of his thinking on this subject. He emphasized that it is not simply the non-unitary nature of the collapse that leads one to the conclusion that it cannot be described by the laws of physics, but rather its unpredictability. One could imagine a non-unitary evolution that was nevertheless predictable.

20. Wigner's views changed significantly toward the end of his life, but not, it should be said, in the direction of materialism. Wigner became convinced, on the basis of some work by H. Dieter Zeh, that it was not realistic to talk about macroscopic systems as being "isolated," and that therefore the density matrix of any macroscopic object cannot evolve in a unitary way. Consequently, he concluded that in the process of measurement the quantum theoretical description breaks down at a point before the consciousness of the observer is affected, and in particular at the point when a macroscopic object (such as a measuring apparatus) becomes involved. He therefore concluded that the collapse

of the wavefunction happens as a result of the interaction of quantum systems with macroscopic systems, rather than as a result of the knowledge of the outcome entering the consciousness of the observer. He felt, however, that if anything these new considerations put the phenomena of consciousness even farther outside the domain of validity of present quantum theory. I believe that Peierls's way of framing the argument is unaffected by these considerations.

21. Stapp, *Mind, Matter, and Quantum Mechanics.*

22. Squires, *Conscious Mind in the Physical World.*

23. Davies and Brown, *The Ghost in the Atom,* 64.

24. W. Heisenberg, *Daedalus* 87 (1958): 99, quoted by Wigner, *Symmetries and Reflections,* 172.

25. Davies and Brown, *The Ghost in the Atom,* 47.

26. Squires, *Conscious Mind in the Physical World,* 192.

27. Peierls, "Observation in Quantum Mechanics and the 'Collapse of the Wavefunction,'" 194.

28. Wigner, *Symmetries and Reflections,* 176.

Chapter 25.   Alternatives to Traditional Quantum Theory

1. Hume's prescription for how to carry on in practical life despite holding radically skeptical philosophical views was "carelessness and inattention," that is, to ignore in practice the implications of one's theory. Hume, *A Treatise of Human Nature,* 144.

2. "[The] debate [on the conceptual foundations of quantum theory] has been argued at various levels since quantum mechanics was first formulated, but the standard interpretation, the so-called Copenhagen interpretation, has stood the test of time and is widely accepted. Alternative formulations appear ultimately to yield the same description. The major advance in the past thirty years was the analysis of John Bell of possible incompleteness in the quantum description. He proposed an experimental test that was carried out and which provided dramatic evidence for the standard formulation. So unless the debate leads to new experimental tests or provides a simpler description of known facts, it will continue to lie outside of the mainstream of physics." Daniel Kleppner, "Physics and Common Nonsense," in *The Flight from Science and Reason,* ed. Paul R. Gross, Norman Levitt, and Martin W. Lewis (Baltimore: Johns Hopkins University Press, 1996), 129–30.

3. T. Banks, "The State of Matrix Theory," 341.

4. D. Bohm, "A Suggested Interpretation of the Quantum Theory in Terms of 'Hidden Variables,'" parts I and II, *Physical Review* 85 (1952), 166–79, 180–93; D. Dürr, S. Goldstein, and N. Zanghi, "Quantum Equilibrium and the Origin of Quantum Uncertainty," *Journal of Statistical Physics* 67 (1992): 843–907, and references therein.

5. Albert Einstein, *The Meaning of Relativity* (Princeton: Princeton University Press, 1945), 55–56.

6. H. Everett, *Review of Modern Physics* 29 (1957): 454–62; *The Many-Worlds Interpretation of Quantum Mechanics,* ed. Bryce S. de Witt and Neill Graham (Princeton: Princeton University Press, 1973).

7. See the analysis of one such attempt in Euan Squires, "On an Alleged 'Proof' of the Quantum Probability Law," *Physics Letters* A145 (1990): 67.

8. See, for example, J. S. Bell, "Quantum Mechanics for Cosmologists," in *Quantum Gravity*, ed. Christopher J. Isham, Roger Penrose, and Dennis W. Sciama, vol. 2 (Oxford: Oxford University Press, 1981); and David Bohm and Basil J. Hiley, *The Undivided Universe: An Ontological Interpretation of Quantum Theory* (London: Routledge, 1993).

Chapter 26.   Is a Pattern Emerging?

1. Penrose, *Shadows of the Mind*, 28.
2. Chalmers, *The Conscious Mind*, 128.
3. Ibid., 168, 379.
4. Nagel and Newman, *Gödel's Proof*, 101.
5. Wigner, *Symmetries and Reflections*, 222.
6. Mark Steiner, *The Applicability of Mathematics as a Philosophical Problem* (Cambridge, Mass.: Harvard University Press, 1998), 13–14.

Appendix A.   God, Time, and Creation

1. Many authors express the view that the notion of causality is intrinsically bound up with the notion of time, and that it is a metaphysical necessity that all causes precede their effects in time. However, it would be more accurate to say that our notions of time are bound up with our notions of causality. A careful consideration of the problem of the "arrow of time" in physics leads to the conclusion that the past/future distinction is rooted in the causal structure of the world. The notion of causality would seem to be the more general and fundamental one.

2. See, for example, Trinh Xuan Thuan, *The Secret Melody: And Man Created the Universe*, trans. Storm Dunlap (Oxford: Oxford University Press, 1995), 242–43.

3. St. Thomas Aquinas, *Summa Contra Gentiles*, bk. I, trans. Anton Pegis, F.R.S.C. (Notre Dame, Ind.: University of Notre Dame Press, 1975), 86–95.

4. Etienne Gilson, *The Christian Philosophy of St. Thomas Aquinas*, trans. L. K. Shook, C.S.B. (Notre Dame, Ind.: University of Notre Dame Press, 1994), pt. I, ch. 3.

5. Ibid., 68.

6. St. Thomas Aquinas, *Summa Contra Gentiles*, bk. II, 112.

7. Ibid., 114.

8. Ibid., 113. Unfortunately, the examples that Aquinas used to illustrate the concept of a cause that is simultaneous with its effect were based on naive observations of the physical world or on Aristotelian physics. I have tried to give a better example in the text.

9. Of course, if a mass were to be suddenly moved, the gravitational field at some distance from that mass would only change in response a finite time later. Information cannot propagate faster than the speed of light.

10. Gilson, *The Christian Philosophy of St. Thomas Aquinas*, 67–68.

11. I am not concerned in this appendix to argue for the validity of the First Mover proof. My purpose is only to clarify its meaning. For a claim by a theist that the First

Mover proof is invalid, see Mortimer J. Adler, *How to Think about God,* (New York: Macmillan, 1980). For a proof of the existence of God by a recent philosopher that is essentially along the lines laid out by St. Thomas Aquinas, see Lonergan, *Insight: A Study of Human Understanding,* ch. XIX. Lonergan considered the famous five proofs of St. Thomas to be valid and to be contained in his own proof. (To understand Lonergan's proof, one has to have worked through the preceding parts of his book. Though Lonergan's book is by no means easy reading, it contains formulation of traditional Thomistic philosophy in terms that I think would make sense to many contemporary physicists. They would also find its epistemology reasonable and, given the recent attacks on the objectivity of scientific knowledge, refreshing.)

12. Richard Swinburne, *The Existence of God* (Oxford: Clarendon Press, 1979), 122, quoted in Davies, *God and the New Physics,* 42.

13. Davies, *God and the New Physics,* 42.

14. Don Page, "Hawking's Timely Story," *Nature* 332 (21 April 1988), 743.

15. St. Thomas Aquinas, *Summa Contra Gentiles,* bk. II, 46.

16. St. Augustine, *Confessions,* 287.

17. Lonergan, *Insight: A Study of Human Understanding,* 639–77, especially 676–77.

18. See, for example, Dawkins, *Climbing Mount Improbable,* 77. Dawkins argues that God, as creator of the universe, would have to be more complicated than the universe, and therefore would himself stand in need of being explained.

19. See, for example, Thuan, *The Secret Melody,* 241.

20. Stenger, *Not by Design,* 7. See also Ferguson, *The Fire in the Equations,* 137–45.

21. Quoted in Davies, *God and the New Physics,* 122.

22. Lonergan, *Insight: A Study of Human Understanding,* 676–77.

23. St. Augustine, *Confessions,* bk. 11, 283.

24. Genesis Rabbah Parashiyyot I.I.2 (citing Prov. 8:22), quoted and discussed in Neusner, *Confronting Genesis,* 15, 26.

25. The divine Wisdom is personified in several books of the Old Testament, notably Proverbs, Wisdom, Baruch, and Sirach.

26. St. Augustine, *Confessions,* 284.

27. Neusner, *Confronting Genesis,* 15.

Appendix B. Attempts to Explain the Beginning Scientifically

1. Stephen Hawking, *A Brief History of Time,* 174.

Appendix C. Gödel's Theorem

1. For an introduction to the philosophy of mathematics, see Stewart Shapiro, *Thinking about Mathematics: The Philosophy of Mathematics* (Oxford: Oxford University Press, 2000).

2. Newman and Nagel, *Gödel's Proof.*

Index

About the Author

STEPHEN M. BARR is a professor of theoretical particle physics at the Bartol Research Institute of the University of Delaware.